Third Bioprocessing and Biomanufacturing Symposium (BBS 2019)

Think Ideas, Talk Innovation, Transform Industries

IOP Conference Series: Materials Science and Engineering
Volume 716

Lumut, Malaysia
9 - 10 April 2019

Editors:

Mohd Asyraf Kassim
Mohd Shamzi Mohamed

ISBN: 978-1-7138-2575-3
ISSN: 1757-8981

Printed from e-media with permission by:

Curran Associates, Inc.
57 Morehouse Lane
Red Hook, NY 12571

Some format issues inherent in the e-media version may also appear in this print version.

Institute of Physics
Dirac House, Temple Back
Bristol BS1 6BE UK

Phone: 44 1 17 929 7481
Fax: 44 1 17 920 0979

techtracking@iop.org

Additional copies of this publication are available from:

Curran Associates, Inc.
57 Morehouse Lane
Red Hook, NY 12571 USA
Phone: 845-758-0400
Fax: 845-758-2633
Email: curran@proceedings.com
Web: www.proceedings.com

TABLE OF CONTENTS

FERMENTATION TECHNOLOGY

BIOMANUFACTURING OF BIOLOGICAL CELLS AND METABOLITE

DOWNSTREAM PROCESSING TECHNOLOGY

ENZYMATIC AND MICROBIAL PROCESSING

PRODUCT BIOCHEMISTRY

IOP Conf. Series: Materials Science and Engineering **716** (2020) 011001 doi:10.1088/1757-899X/716/1/011001

THE THIRD BIOPROCESSING & BIOMANUFACTURING SYMPOSIUM (BBS) 2019

"Think Ideas, Talk Innovation, Transform Industries"

9-10th April 2019

Swiss Garden Beach Resort,
Lumut, Perak
Malaysia

IOP Conf. Series: Materials Science and Engineering **716** (2020) 011001 doi:10.1088/1757-899X/716/1/011001

Preface

The Third Bioprocessing and Biomanufacturing Symposium (BBS) 2019 was held at Swiss Garden Bearch Resort, Lumut Perak on 9-10th April 2019 organized by Bioprocess Technology Division, School of Industrial Technology, Universiti Sains Malaysia and Bioprocessing and Biomanufacturing Research Center, Universiti Putra Malaysia (UPM) and Department of Bioprocess Technology, Faculty of Biotechnology and Biomolecular Sciences, UPM. This is a biennial conference series that bringing together most of the bioprocess researchers, academics and professionals from all in the country.

Bioprocessing technology is growing in significance and highly relevant in the context of exploring new knowledge and integration of various interdecipline to generate new innovation in various field such as bio-therapeutic, environmental management and sustainable bioprocess development.

This symposium provides a unique platform for leading experts and researchers from academic and industry sector to promote the exchange of information, ideas and intellectual to catapult trans-disciplinary research, collaborations and networking among experts in various areas of bioprocessing and biomanufacturing technology, particularly with the advancement of the various fermentation instruments and development of new method or approach which enable us to produce a wide range of value added products.

The symposium comprised of 25 highly informative technical papers that comprehensively cover latest finding of the sub-themes such as fermentation technology, biomanufacturing of biological cells and metabolite, enzyme and microbial bioprocessing, downstream processing technology and product biochemistry. On this occasion, a total of 4 distinguished plenary speaker from academia and industry had delivered their outstanding research and sharing the current scenario in bioprocessing and biomanufacturing industry.

Lastly, we sincerely hope that this BBS2019 is a significant step forward in the effort to improve and overcome the current challenges in the bioprocessing and biomanufacturing. Finally the board of editors would like to thank all the reviewers contribution to improve the quality of the papers by providing constructive comment and suggestions. On the other hand, we would like to thank the authors who has devoted a subtantial amount of work to make this high quality BBS2019 proceeding success.

Dr. Mohd Asyraf Kassim
Chairman
The Third Bioprocessing and Biomanufacturing Symposium (BBS) 2019

IOP Conf. Series: Materials Science and Engineering **716** (2020) 011001 doi:10.1088/1757-899X/716/1/011001

Committee Member

Advisor:	Prof Dr. Arbakariya Arif Assoc. Prof. Dr. Rosma Ahmad Prof. Dr. Norli Ismail
Chairman:	Dr Mohd Asyraf Kassim
Vice-Chairman:	Dr. Mohd Shamzi Mohamed
Secretary:	Dr. Mohamad Hafizi Abu Bakar
Assistant Secretary:	Pn. Juliana Jaafar
Finance/ Sponsorship:	Dr. Tan Joo Shun Mr. Sobri Mohd Akhir
Treasurer:	Dr Lee Chee Keong
Assistant Treasurer:	Mdm Liyana Ithnin
Scientific:	Dr. Mohamad Hafizi Abu Bakar Dr. Murni Halim Dr. Zetty Norhana Balia Yusoff Dr. Siti Efliza Ashari Dr. Nur Adeel Yasid Assoc. Prof. Dr. Nor'Aini Abdul Rahman Dr. Fazlie Wong Faizal Wong Dr. Mohd Ezuan Khayat
Publicity/ICT:	Dr Noor Aziah Serri Ms Wan Zafira Ezza Wan Zakaria Dr. Mohamad Zulfazli Mohd Sobri Dr. Helmi Wasoh
Assistant Publicity/ICT:	En. Muhammad Khairi Abd Aziz
Technical and Transportation:	Dr. Muaz Mohd Zaini Makhtar Dr. Husnul Azan Tajaruddin Postgraduate Students
Social and Protocol:	Prof. Liong Min Tze Dr Siti Baidurah Yusoff Prof. Dr. Rosfarizan Mohamad

IOP Conf. Series: Materials Science and Engineering **716** (2020) 011001 doi:10.1088/1757-899X/716/1/011001

Editorial Board

IOP Conf. Series: Materials Science and Engineering **716** (2020) 011001 doi:10.1088/1757-899X/716/1/011001

Peer review statement

All papers published in this volume of IOP Conference Series: Materials Science and Engineering have been peer reviewed through processes administered by the proceedings Editors. Reviews were conducted by expert referees to the professional and scientific standards expected of a proceedings journal published by IOP Publishing.

Conference Photographs

IOP Conf. Series: Materials Science and Engineering **716** (2020) 011001 doi:10.1088/1757-899X/716/1/011001

IOP Conf. Series: Materials Science and Engineering **716** (2020) 011001 doi:10.1088/1757-899X/716/1/011001

Peer review statement

All papers published in this volume of *IOP Conference Series: Materials Science and Engineering* have been peer reviewed through processes administered by the proceedings Editors. Reviews were conducted by expert referees to the professional and scientific standards expected of a proceedings journal published by IOP Publishing.

IOP Conf. Series: Materials Science and Engineering **716** (2020) 012001 doi:10.1088/1757-899X/716/1/012001

Potentials of sago fibre hydrolysate (SFH) as a sole fermentation media for bioethanol production

S Mohammad[1], D S Awg-Adeni[2*], K B Bujang[2], M Vincent[2], S Baidurah[1]

[1]School of Industrial Technology, Universiti Sains Malaysia, 11800 Minden, Penang, Malaysia
[2]Resource Biotechnology Programme, Faculty of Resource Science and Technology, Universiti Malaysia Sarawak, 94300 Kota Samarahan, Sarawak, Malaysia

Email: adsalwa@unimas.my

Abstract. Sago wastewater which contains starchy fibres from sago starch processing mills is commonly discharged directly to nearby stream thus contribute to serious environmental pollution. Sago fibre which is known to be a local agricultural waste mainly contains residual starch of about (50 – 60 %) together with cellulosic component. These contribute to high carbohydrate contents which suitable to be used as substrate for ethanol production. Initially, sago fibre (SF) was converted into sago fibre hydrolysate (SFH) via enzymatic hydrolysis using commercial enzymes; Liquozyme SC DS and Spirizyme Fuel HS. This study emphasized on batch ethanol fermentation by commercial baker's yeast utilizing 50 g/L and 80 g/L glucose of SFH as the sole fermentation medium. The results indicate that 50 g/L glucose from SFH media is capable of generating maximum ethanol concentration at 20.33 ± 0.15 g/L, with highest glucose consumption efficiency (97.78 %) during 24 hours of fermentation. Similar concentration of bioethanol was obtained in 50 g/L glucose of commercial glucose (CG) media which is at 20.04 ± 0.06 g/L. However, lower ethanol concentration was obtained in both 80 g/L glucose from SFH (13.32 ± 0.12 g/L) and CG (12.98 ± 0. 04 g/L media), respectively. Addition of yeast extract at 3 g/L into 80 g/L SFH as well as CG significantly improve ethanol fermentability (SFH: 41.04 ± 0.04 g/L and CG: 33.96 ± 0.04 g/L). Based on statistical analyses, 50 g/L glucose of SFH media exhibit the highest ethanol yield (0.42 ± 0.003 g/g) and highest fermentation efficiency (81.35 ± 0.572 %) compared to 80 g/L glucose (0.24 ± 0.008 g/g; 46.65 ± 1.50 %). Conclusively, this study demonstrated that glucose in SFH was metabolized efficiently by commercial baker's yeast during ethanol fermentation, thus suggesting the capability of SFH to be a feasible and alternative substrate with less expensive nitrogen source for the renewable bioethanol production.

1. Introduction

The waste to wealth concept research has been practiced actively over the last few years, which widely focusing on the bioethanol production utilizing starchy lignocellulosic residue. Bioethanol (C_2H_5OH) or ethyl alcohol is a promising alternative source which is both renewable and environmental friendly [1]. Over the years, bioethanol production has increased dramatically due to high demand on green biofuel and crude oil shortage [2]. At present, global bioethanol production has reached about 5,340 million litres in 2017 and slightly increased to 5,380 million litres in 2018. This demand is met by using first generation bioethanol crops such as sugarcane and corn [3]. The research into new and alternative carbon sources such as industrial by- products and agricultural residues for uses in bioethanol fermentation is needed to sustain for the successful development of fermentation industries against traditional grain-based processes. Thus, a renewable raw material, which is locally abundant, can be considered as an excellent feedstock for bioethanol production by virtue of high carbohydrate content, would be more preferable. In Malaysia, Sarawak is recognized as having the largest sago-growing areas, which is currently the world's biggest exporter of sago starch, exporting annually about 44,000 tonnes of starch mainly to peninsular Malaysia, Japan, Singapore and other countries [4]. It was estimated that starch contributed to about 15-16% (dry weight) of sago palm log [5]. Hence, for every 1 ton of sago starch produced, 6.5 ton palm's log was processed that leads to 20 ton of wastewater and 1 ton of sago fibre were discharged [5]. Approximately, 60,000 tonnes of sago fibre are produced daily from eight sago starch processing mills in Mukah [6]. Currently, these wastes are being washed off into nearby rivers or deposited in the factory's compound daily. This condition potentially deteriorates the rivers and the environment [7]. Sago fibre, which is free and abundant - is derived as a starchy lignocellulosic by-product from extraction of starch from the pith of the sago palm. On a dry basis, it contains 58 % residual starch together with high content of cellulosic components [8]. Previously, the sugar hydrolysate obtained from enzymatic hydrolysis of sago starch has been revealed by our group to be feasible as a cheap substrate for ethanol fermentation using *Zymomonas mobilis* [9]. Later, additional of trace elements as well as nitrogen and phosphorus would enhances the production of ethanol from sago fibre [10]. In any ethanol fermentation process, the greater substrate load would lead to increase in ethanol concentration, and improve the efficiency of downstream processing. Based on previous study, hydrolysate of sago fibre cannot be promoted as the sole carbon source if higher concentration of glucose (>100 g/L) was employed [11]. Thus, lower concentration of substrate was implemented in order to obtain maximum ethanol concentration. This projects attempts to study the potentials of sago fibre hydrolysate (SFH) as the sole carbon source in bioethanol production utilizing 50 g/L and 80 g/L of initial glucose concentration by commercial baker's yeast.

2. Materials and Methods

2.1 Sago fibre

Sago fibre samples is obtained from the Ubom Sago Mill Sdn. Bhd. in Mukah, Sarawak in dried condition .The fibre was ground to pass 1 mm screen prior to produce sago fibre powder (SFP); to be used as raw material for glucose production and compositional analysis.

2.2 Enzymatic hydrolysis of sago fibre powder (SFP)

A suspension of sago fibre, 7 % (w/v) was prepared in 0.1 M monopotassium phosphate (KH_2PO_4) buffer solutions at pH 4. The suspension was then boiled and gelatinized at 85 - 90 ºC for about 30 minutes. After gelatinization, commercial enzymes of Liquozyme® SC DS and Spirizyme® Fuel HS (Novozymes, Denmark) were added into the suspension during liquefaction and saccharification stage. Upon completion of the hydrolysis process, the slurry obtained was filtered from its solid residual fibre, producing sago fibre hydrolysate (SFH). In order to maximize glucose recovery, recycle method was adopted in this study [10].

2.3 Inoculum preparation

An instant dry baker's yeast, labelled as Mauripan; *Saccharomyces cerevisiae* was used throughout this study. The yeast which was cultured on potato dextrose agar was transferred into sterile 100 mL inoculum media containing 20 g/L glucose and 5 g/L yeast extract. The pH for inoculums was adjusted at 5.5 before being incubated at 30 ºC for 9 - 10 hours. Once the optical density (OD) of yeast cell reached to 0.5 to 0.6, which indicates log phase of the yeast growth profiles, the baker's yeast was developed and yeast was ready to be harvested. 10 mL of inoculum media was then removed and centrifuged in order to obtain the yeast pellet which then ready to be transferred into sterile fermentation media.

2.4 Fermentation media and its process

SFH with initial glucose concentration of 50 g/L and 80 g/L was used as the main medium for ethanol fermentation and commercial glucose (CG) was used as a control throughout this study. The fermentation media was supplemented with nitrogen source; yeast extract, (3 g/L) [10]. The fermentation process was conducted via batch system, working volume of 100 mL shake flask, agitated at 100 rpm, 30℃ for 24 hours. In every 4 hours interval, the samples withdrawn were centrifuged at 10, 000 rpm for about 10 minutes at 4℃. The brown clear supernatant was then used for determination of ethanol concentration and glucose consumption by High Performance Liquid Chromatography (HPLC) system (Shimadzu, Kyoto, Japan).

2.5 Analytical procedures

Moisture content of sago fibre was determined using standard oven drying method at 105 ^{o}C to constant weight [12]. Analyses of crude fibre, crude fat, total ash and energy were sent to the Veterinary Diagnosis Laboratory, Sarawak. The residual starch content of sago fibre was measured by iodine starch colorimetric method [13]. Total reducing sugars was determined by using DNS method [14], whereas glucose and ethanol were analysed by using HPLC system (Shimadzu, Kyoto, Japan) equipped with a Refractive Index (RID-10A) detector with a column of prominence CTO - 20A (Aminex HPX – 87H column, 150 x 7.8 mm, BioRad Laboratories, Inc). Mobile phase used was 5 mM H_2SO_4 with a flow rate set at 0.8 mL/min and column temperature was fixed at 60 °C.

2.6 Data analysis

The data obtained from a series of fermentation process was compiled and analysed for ethanol yield (Y, g/g), ethanol fermentation efficiency (E_f, %) and ethanol volumetric productivity (Q_p, g/L/hr). Ethanol yield was calculated as the actual ethanol produced and expressed as gram ethanol per gram of glucose consumed (g/g). Theoretical yield of maximum ethanol was based on value 0.51. The fermentation efficiency (%) was calculated based on ethanol yield divided by theoretical yield of 0.51. Ethanol volumetric productivity was calculated as the maximum ethanol concentration obtained divide by the fermentation time (hrs).

2.7 Statistical analyses

Data obtained from all of experiments were compared and analysed using a One Way ANOVA. LEAD Technologies, MINITAB Version 17.1.0 was used to perform the statistical analyses.

3. Results and Discussion

3.1 Characterization of the sago fibre powder (SFP)

The composition of sago fibre powder is presented in Table 1. It was observed that the amount of residual starch in sago fibre was highest from CL Nee (Sibu) at 70.00 %, followed by Hup Guan (Johor, 65.70 %) and Ubom (Mukah, 59.00 – 60.00 %).

Table 1. Compositional analyses of the sago fibre powder (as percentage dry weight)

	This study (2017)	Abd - Aziz (2002)	Kumoro *et al.* (2008)	Awg - Adeni *et al.* (2013)
Sources of sago fibre	Ubom, Mukah	Hup Guan, Johor	CL Nee, Sibu	Herdsen, Pusa
Residual starch	59.00 – 60.00	65.70	70.00	30.00 – 45.00
Moisture content	10.43	5.91	6.10	5.00 – 7.00
Crude Ash	3.35	4.10	4.16	3.00 – 4.00
Crude Fibre	22.94	14.80	18.76	30.00 – 35.00

Note: Apart from Hup Guan in Johor, all other factories are in Sarawak.

In this study, the residual starch content in SFP was 59.00 – 60.00 % (on a dry weight basis), which almost comparable to the starch contents reported by previous studies, which is in the range of 30 to 70 % [10, 15, 16]. The amount of residual starch (59.00 – 60.00 %) observed in sago fibre was quite high as compared to other studies due to several factors. The sago fibre is sourced from different sago mills, thus different methods or degree of starch extraction together with different facilities contribute to different quality of sago extraction process [10]. Most large sago mills have between 10 to 12 extractors which efficiently separate dissolved starch from the fibres. Smaller mills would have lesser number of extractors, which leads to higher percentage of residual starch in the effluent and attached to the fibres. Higher percentages of residual starch in sago fibre (SF) exhibited a less efficient starch extraction procedure in these factories, which leads to profit loss, disposal of large quantities of starch into the river and eventually leads to environmental pollution. This demands prompt and effective utilization of the sago fibre to minimize pollution and processing it into beneficial products to generate side-incomes for these sago mills. Moreover, the amount of residual starch and other composition including free sugars differ from every sago processing mill, depends on the effectiveness of the starch harvesting, extraction, filtration and sedimentation system used [17]. Thus, the amount of residual starch that still trapped within sago fibre would also being influenced by all the differences practices or techniques applied in sago factories.

3.2 Enzymatic hydrolysis of residual starch in SF for glucose production

Theoretically, in any chemical reaction, higher substrate concentration would have resulted in higher concentration of the end-product. Based on this study, recycled method was adopted in order to maximize the glucose production from SF. In every cycle, 7 % (w/v) suspension was prepared, thus total substrate load was accounted to be 21 % (w/v). The conversion of glucose was expressed in terms of glucose concentration (g/L) by using HPLC analysis. Table 2 represents the glucose concentrations (g/L) of sago fibre hydrolysate (SFH) for three-cycle enzymatic hydrolysis process. The glucose concentration after cycle I of enzymatic hydrolysis was 46.24 g/L. When SFH from cycle I was added with new substrate load for cycle II, 100.10 g/L glucose was generated. At the end of cycle III, maximum amount of glucose obtained was increased up to 149.36 g/L.

Table 2. Glucose production of three-cycle enzymatic hydrolysis of SFP

Cycle	Glucose Concentration, (g/L)
I	$46.24^{c} \pm 2.27$
II	$100.10^{b} \pm 2.96$
III	$149.36^{a} \pm 1.53$

*Means with the same letter are not significant at $p<0.05$.

From the results, 7 % (w/v) SFP shows significantly higher glucose concentration as the total substrate load accounted, 21 % (w/v) upon completion of three-cycle enzymatic hydrolysis process. It was revealed that conducting hydrolysis process was much easier for cycle II and cycle III due to better solubilisation properties [10]. Addition of enzyme in subsequent hydrolysis cycle would maintain the substrate and enzyme ratio [18]. Repeating and boiling during enzymatic hydrolysis process will assists the starch granules dispersion, thus contribute to higher recovery of starch into glucose [19].

3.3 SFH as the sole fermentation medium for bioethanol production

In this study, SFH (without addition of any nitrogen source) was used as the sole fermentation medium for generating alcohol utilizing 50 g/L and 80 g/L of initial glucose concentration, compared to commercial glucose. Figure 1 represents the fermentation profiles, indicating glucose consumption in SFH and CG media by baker's yeast during 24 until 48 hours of fermentation.

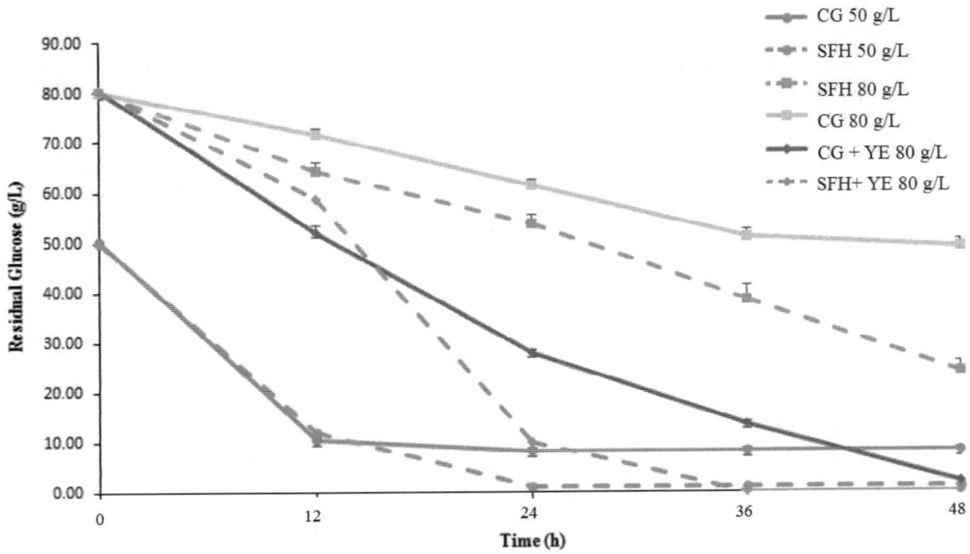

Figure 1. The glucose consumption profiles in SFH fermentation by commercial baker's yeast (CBY) under 50 g/L and 80 g/L initial glucose concentration.

Based on the fermentation profiles, it was observed that, 50 g/L of glucose in SFH media was almost consumed by the commercial baker's yeast (CBY) during 24 hours of fermentation, with remaining glucose at 1.11 ± 0.01 g/L. Thus, the total glucose consumption in 50 g/L SFH media was 97.78 %, which indicates the fastest glucose consumption, compared to 50 g/L glucose in CG media, with total glucose consumption of 83.81 %, at 24 hours of fermentation. In contrast to 80 g/L of initial glucose concentration, it was observed that, 80 g/L of glucose in SFH media was not completely consumed by the yeast during 48 hours of fermentation, with remaining glucose of 24.155 ± 1.40 g/L, thus indicates 69.83 % of glucose consumption. Similar trend was also observed under 80 g/L of glucose in CG media. The remaining glucose was at (48.99 ± 1.40 g/L), with lower total glucose consumption of 38.75 %. In this fermentation condition, 80 g/L glucose in SFH media demonstrated slightly better performance on glucose consumption rather than in 80 g/L glucose of CG media.

Figure 2 represent the fermentation profiles, indicating ethanol production in SFH fermentation by CBY during 24 until 48 hours of fermentation. Based on the results obtained, it was observed that, 50 g/L glucose from SFH media can generate maximum ethanol during 24 hours of fermentation, which at 20.33 ± 0.15 g/L. Similar ethanol concentration was obtained in 50 g/L glucose of CG media, which is at 20.04 ± 0.06 g/L. However, in 80 g/L glucose of SFH, lower ethanol was produced (13.32 ± 0.12 g/L, 7 % lower) than in 50 g/L SFH media. Similar ethanol profile was obtained under 80 g/L glucose in CG fermentation, which is about 12.98 ± 0.04 g/L. Based on the ethanol production profiles, higher and maximum concentration of ethanol was achieved when lower initial glucose concentration (50 g/L) was introduced, whereas lower concentration of ethanol was generated when higher initial glucose concentration was introduced (80 g/L). In this condition, the yeast might be unprotected to severe conditions such as high osmotic pressure, which increased the by-product inhibition and nutritional limitations [20]. Normally, inadequate of nitrogen source may also cause inhibition towards the yeast growth, thus reduce the ethanol conversion, as well as ethanol yield [21, 22].

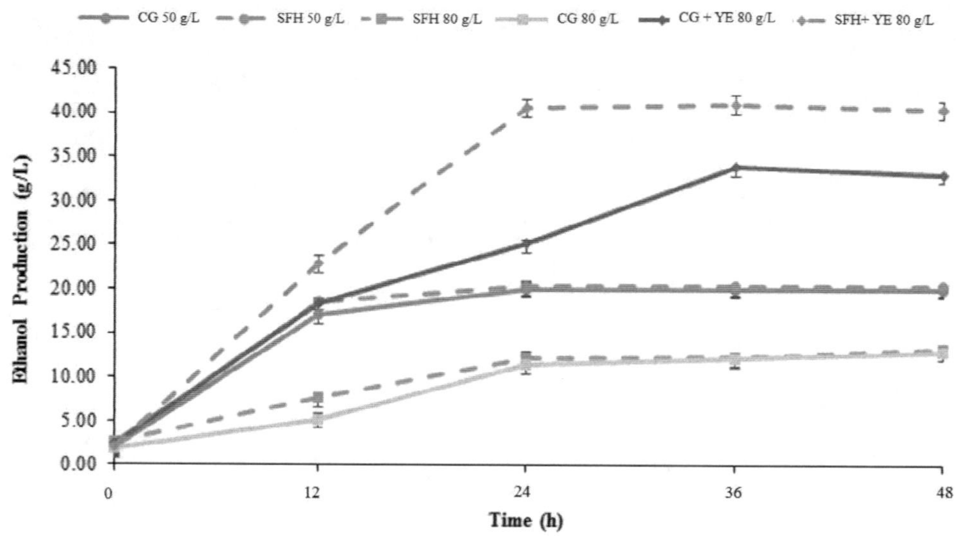

Figure 2. The ethanol production profiles in SFH fermentation by baker's yeast under 50 g/L and 80 g/L initial glucose concentrations.

Table 3 represent the fermentation performance of SFH media utilizing 50 g/L and 80 g/L glucose as the sole carbon source for production of bioethanol. Based on the kinetic parameters, 50 g/L of initial glucose concentration in SFH media shows the highest glucose consumption efficiency (97.78 %), higher fermentation efficiency (81.35 ± 0.572 %), with respective ethanol yield of 0.42 ± 0.003 g/g, indicating that CBY was well tolerates with SFH as the sole carbon source for ethanol production. However, when glucose concentration increased up to 80 g/L either in SFH or CG media, the ethanol fermentability decreased tremendously. Hence, selected yeast extract had been added into the fermentation media in order to improve ethanol fermentation process. Throughout the trial of adding yeast extract in fermentation media containing 80 g/L of initial glucose concentration, improved ethanol concentration were obtained in SFH (41.04 ± 0.04 g/L) and CG (33.96 ± 0.04 g/L). Both media able to generate maximum ethanol concentration during 36 hours of fermentation, much earlier than expected time for the fermentation to complete. This result indicates that glucose in SFH media (amended with YE) was fully metabolized by the CBY after 24 hours of fermentation, with 100 % glucose consumption efficiency. However, in 80 g/L glucose of CG media, the glucose was almost utilized during 48 hours of fermentation, indicates 97.33 % of glucose consumption efficiency, slightly slower compared in 80 g/L glucose in SFH media.

Interestingly, both ethanol yield ($Y_{p/s}$, g/g) and fermentation efficiency (E_f, %) were improved from 0.24 ± 0.01 g/g to 0.51 ± 0.002 g/g and 46.65 ± 1.50 % to 99.71 ± 0.46 % in 80 g/L SFH media only and 80 g/L SFH media (amended with YE), respectively. Clearly, the addition of YE as nitrogen source into hydrolysate with 80 g/L glucose, significantly assisting in fermentation process, thus enhanced the ethanol fermentability towards the CBY. In this fermentation condition, once higher initial glucose (>50 g/L) was introduced into the same process, both ethanol yield and fermentation efficiency shows slight decline on their value, although the fermentation time was prolonged in order to achieve maximum ethanol concentration. Usually, the extension of fermentation time is a common practice in high gravity (HG) or very high gravity (VHG) fermentation, which also could decrease the ethanol productivity correspondingly [23]. Supplementation of correct amount and sufficient nitrogen

IOP Conf. Series: Materials Science and Engineering **716** (2020) 012001 doi:10.1088/1757-899X/716/1/012001

sources are crucial in order to increase the ethanol yield, as well as to enhance the fermentation efficiency and ethanol volumetric productivity of the fermentation process.

Table 3. Kinetic parameters of ethanol fermentation utilizing SFH (with addition of yeast extract) at initial glucose of 50 g/L and 80 g/L

Kinetic Parameters	**50 g/L**		**80 g/L**			
	SFH	**CG**	**SFH**	**CG**	**SFH + YE**	**CG+ YE**
Initial Glucose (g/L)	50.00 ± 0.00	50.28 ± 0.14	80.06 ± 0.04	79.98 ± 0.08	80.54 ± 0.44	80.51 ± 0.41
Glucose Consumption Efficiency (%)	97.78	83.81	69.83	38.75	100	83.24
Ethanol (g/L)	20.33 ± 0.15	20.04 ± 0.06	13.32 ± 0.09	12.98 ± 0.05	41.04 ± 0.04	33.96 ± 0.04
$Y_{p/s}$ (g/g)	0.42 ± 0.00	0.48 ± 0.01	0.24 ± 0.01	0.42 ± 0.01	0.51 ± 0.00	0.51 ± 0.00
Q_p (g/L/h)	0.85 ± 0.01	0.84 ± 0.00	0.28 ± 0.00	0.27 ± 0.00	1.14 ± 0.00	0.94 ± 0.00
E_f (%)	81.35 ± 0.57	93.07± 0.10	46.65 ± 1.50	82.00 ± 2.25	99.71 ± 0.46	99.15 ± 0.11

Notes: Parameters studied: $Y_{p/s}$ ethanol yield (g/g), Q_P ethanol volumetric productivity (g/L/h), E_f fermentation efficiency (%). All data presented are the average of duplicate experiments with standard errors. Theoretical ethanol yield is 0.511 g/g glucose.

4. Conclusion

Sago fibre hydrolysate (SFH) obtained via enzymatic hydrolysis of sago fibre powder has been discovered to be an alternative substrate for producing bioethanol using commercial baker's yeast. This study suggested that ethanol fermentation in SFH as a sole fermentation media was limited at 50 g/L initial glucose. The addition of yeast extract however able to boost ethanol fermentability when initial glucose had been increased up to 80 g/L.

Acknowledgement

The authors would like to gratefully acknowledges Universiti Malaysia Sarawak, Tun Openg Research Chair (F07 (ORC29)/1235/2015 (16)) and Universiti Sains Malaysia, Research University Grant (RUI) (1001/PTEKIND/8011022) for the financial support.

References

[1] Vincent M, Jabang E, Muhamad Nur N, Esut E, Unting L B and Awg-Adeni D S 2015 Simultaneous co-saccharification and fermentation of sago hampas for bioethanol production. *Agric. Eng. Int. CIGR J.* **17**(2) 160-67

[2] Saifuddin N, Zhan L W and Ning K K X 2011 Heat-modelling of microwave assisted epoxidation of palm acid oil. *Am. J. Appl. Sci.* **8**: 217-29

[3] Puri M, Abraham R E, Colin I, and Barrow J 2012 Biofuel production: prospects, challenges and feedstock in Australia. *Renew. Sust. Energ. Rev.* **16**(8): 6022-31

[4] Awg-Adeni D S, Abd-Aziz S, Bujang K B and Hassan M A 2010 Bioconversion of sago residue into value-added products. *Afr. J. Biotechnology.* **9**(14): 2016-21

[5] Bujang K B, Apun K and Salleh M A 1996 A study in the production and bioconversion of sago waste. Proceeding 6th International Sago Symposium. In: Jose, C. and Rasyad, R. (Eds.) Sago: The Future Source of Food and Feed. Universitas Riau, Pekan Baru, Sumatra. Republic of Indonesia. 9-12th December, 1996. pp: 193-200

[6] Manan D M A, Chie R and Rumie A M 2003 Sago starch technology: the Sarawak experience. CRAUN Bulletin. Sarawak, Malaysia: CRAUN Research Sdn. Bhd.

[7] Bujang K B 2015 Inaugural lecture. Rejuvenation of an old crop: roles of sago in the food and energy industries. Universiti Malaysia Sarawak (UNIMAS). ISBN: 978-967-5527-83-8

[8] Linggang S, Phang L Y, Wasoh M H and Abd-Aziz S 2012 Sago pith residue as an alternative cheap substrate for fermentable sugars production. *Appl. Biochem. Biotechnol.* **167**: 122-31

[9] Bujang K B, Awg-Adeni D S and lshizaki A 2000 Effects of pH on production of ethanol from hydrolysed sago starch utilising *zymomonas mobilis. Biotechnol. Sustain. Util. Biol. Resour. Trop.* **14**: 20-26

[10] Awg-Adeni D S, Bujang K B, Hassan M A and Abd-Aziz S 2013 Recovery of glucose from residual starch of sago hampas for bioethanol production. *Biomed. Res. Int.* Article ID 935852. DOI:10.1155/2013/935852

[11] Awg-Adeni D S 2015 Bioethanol production from residual starch of sago (*metroxylon sagu* rottb.) hampas. Thesis Doctor of Philosophy. Faculty of Biotechnology and Biomolecular Sciences. Universiti Putra Malaysia, Serdang, Malaysia

[12] AOAC 1990 *Official Method of Analyses*, *Association of Official Analytical Chemist* ed 15 (Washington DC-USA)

[13] Nakamura L K 1981 *Lactobacillus amylovorus*, a new starch-hydrolyzing species from cattle waste corn fermentation. *Int. J. Syst. Bacteriol.***31**: 56-63

[14] Miller G L 1959 Use of dinitrosalicylic acid reagent for determination of reducing sugar. *Anal. Chem.* **31:** 426-28

[15] Abd-Aziz S 2002 Sago starch and its utilization. *J. Biosci. Bioeng.* **94** (6): 526-29

[16] Kumoro A C, Ngoh G C, Hasan M, Ong C H and Teoh E C 2008 Conversion of fibrous sago *(metroxylon sagu)* waste into fermentable sugar via acid and enzymatic hydrolysis. *Asian J. Sci. Res.* **1**: 412-20

[17] Janggu U and Bujang K B 2009 Maximizing sugar production from enzymatic hydrolysis of sago fibre for ethanol fermentation. 1st ASEAN Sago Symposium. Riverside Majestic Hotel, Kuching. 29-30th October, 2009

[18] Modenbach A A and Nokes S E 2013 Enzymatic hydrolysis of biomass at high-solid loadings-a review. *Biomass Bioenerg.* **56**: 526-44

[19] Sajilata M G, Singhal R S and Kulkarni P R 2006 Resistant starch-a review. *Compr. Rev. Food Sci. Saf.* **5**(1): 1-17

[20] Chunkeng H, Qing Q and Peipei G 2011 Medium optimization for improved ethanol production in very high gravity fermentation. *Biotechnol. Bioeng.* **9**: 1017-22

[21] Shen Y, Guo J S, Chen Y P, Zhang H D, Zheng X X, Zhang X M and Bai F W 2012 Application of low- cost algal nitrogen source feeding in fuel ethanol production using high gravity sweet potato medium. *J. Biotechnol.* **160**: 229-35
[22] Phukoetphim N, Salakkam A, Laopaiboon P and Laopaiboon L 2017 Improvement of ethanol production from sweet sorghum juice under batch and fed-batch fermentations: effects of sugar levels, nitrogen supplementation, and feeding regimes. *Electron. J. Biotechn.* **26**: 84-92
[23] Li F, Zhao X Q, Ge X M and Bai F W 2009 An innovative consecutive batch fermentation process for very high gravity ethanol fermentation with self-flocculating yeast. *Appl. Microbiol. Biotechnol.* **84**: 1079-86

IOP Conf. Series: Materials Science and Engineering **716** (2020) 012002 doi:10.1088/1757-899X/716/1/012002

Selective Fermentation and Prebiotic Index of Sago (*Metroxylon sagu*) Resistant Starches Type III by Rat Fecal Cultures

T Zi-Ni and A Rosma[1]
Bioprocess Technology Division, School of Industrial Technology, Universiti Sains Malaysia, 11800 USM, Penang, Malaysia

[1]Email: rosmah@usm.my

Abstract. One of the criteria for prebiotic substance is its selective fermentation by beneficial gut microbes. Previously, sago resistant starch type III (RS_3) produced from *Metroxylon sagu* starch had been proven to resist digestion by gastrointestinal acidity and enzymes. In this research, sago RS_3 was evaluated for its selective fermentation and prebiotic potential by fecal cultures. Sago RS_3 was incubated with fecal microbial cultures for 72 hours at 37°C and samples were measured for microbial growth, prebiotic index, organic acid concentration and β-glucuronidase activity during the fermentation. Commercial prebiotics FOS (fructooligosaccharides) and inulin were used as comparison. Sago RS_3 was selectively fermented by fecal beneficial lactobacilli and bifidobacteria while decreasing the growth of detrimental bacteroides, clostridia and enterobacteria. The prebiotic index of sago RS_3, FOS and inulin was +12.19, +9.45 and +6.82, respectively. The butyric acid molar concentration in media with sago RS_3 was comparable with FOS and inulin, and low activity of β- glucuronidase was detected in medium with sago RS_3. Sago RS_3 exhibited prebiotic characteristics comparable with commercial prebiotics and their potential prebiotic function is worth for further *in vivo* assessment.

1. Introduction.

Human large intestine is heavily populated by numerous and diverse species of microorganism, forming a complex microflora community. Colonic microflora plays a crucial role in maintaining the proper intestinal functions and this influences the host health by impacting the development of immune system, inhibiting the growth of pathogen and regulating metabolic pathways in the host [1]. Hence, colonic microflora must be maintained in a balanced state which predominantly constitutes of health promoting bacteria, for instance, lactobacilli and bifidobacteria. Imbalance in the composition of colonic microflora may be linked to numerous diseases such as colorectal cancer and inflammatory bowel disease [2].

A promising strategy, whereby involving the usage of prebiotic was introduced [3]. A prebiotic is a non-digestible carbohydrate that selectively stimulates the growth and/or activity(ies) of one or a limited number of microbial genus(era)/species in the gut microbiota that confer(s) health benefits to the [4]. A prebiotic ingredient should resist towards the digestions in the upper gastrointestinal tract and intestinal absorption, and be selectively fermented by intestinal microflora associated with beneficial effects [5]. The demand for prebiotic food is growing rapidly and is expected to reach USD$5.5 billion by the year 2020 [6]. Hence, many researches are in the progress of studying various sources of carbohydrate for their prebiotic potential. Oligosaccharides such as fructooligosaccharides, galactooligosaccharides and trans-galactooligosaccharides are commonly available commercial prebiotics [7] whereas resistant starch as prebiotic is not commercially available. Emerging evidences had revealed that resistant starch could be a potential prebiotic [8]. Sago RS_3 is known for its resistance to digestion by gastric acidity and pancreatic enzymes, and resistant to absorption along the upper intestinal tract [9]. With the beneficial effects of resistant starch towards health [8], it is suggested to incorporate RS in the formulation of various food products as many of the commonly available foods contain low RS amount of less than 10 g/100 g [9] whilst recommended daily consumption of RS is 20 g [10]. No information on the advantages of sago RS3 compared to other prebiotics is available.

In this study, sago RS_3 was evaluated for its fermentation and prebiotic index by fecal cultures. The metabolic activities (organic acids production and β-glucuronidase activity) of the diverse species of gut microorganism will be monitored. The changes of these bacteria population allow us to determine the fermentation selectivity of sago RS_3. The fermentation of sago RS_3 by fecal cultures will be compared with commercial prebiotics FOS and inulin.

2. Materials and methods

2.1 Materials.
Native sago starch used to produce sago resistant starch type III (RS_3) was purchased from a local grocery in Kuching, Sarawak, Malaysia. Fructooligosaccharides (Orafti® P95) and inulin (Orafti® GR) were purchased from BENEO-Orafti, Tienen, Belgium. Pullulanase debranching enzyme (Promozyme®D2) was purchased from Novozymes (Bagsvaerd, Denmark) and used upon receipt. Unless otherwise specified, all other chemicals used were of analytical grade and were purchased from Sigma-Aldrich Inc., St. Louis, Missouri, USA.

2.2 Fecal samples.
The fecal samples were collected from eight healthy male Sprague Dawley rats, housed in the Animal Research and Service Centre, Universiti Sains Malaysia. The 3-month old rats were not on antibiotics or probiotics/prebiotics. The fresh feces were pooled and stored in sterile sealed bags and immediately transported to the laboratory and dispensed into sterile 0.17 M phosphate buffered saline [11] at a concentration of 10% (w/v) in an anaerobic condition. The feces were homogenized using a stomacher for 2 min and used as an inoculum for the fermentation experiment after filtering through two layers of muslin cloth.

2.3 Preparation of Sago RS_3.
Sago RS_3 was produced according to the method as described by Zi-Ni et al. [12].

2.4 Fermentation Media.
Basal fermentation medium [13] contained (per L): 2 g peptone, 2 g yeast extract, 2 g $NaHCO_3$, 0.5 g bile salts, 0.5 g L-cysteine.HCl, 0.1 g NaCl, 0.04 g K_2HPO_4, 0.04 g KH_2PO_4, 0.01 g $MgSO_4.7H_2O$, 0.01 g $CaCl_2.6H_2O$, 0.005 g Hemin, 2 mL Tween 80, 10 μL Vitamin K1 and 4 mL Resazurin solution (0.025%, w/w). The fermentation media was adjusted to pH 7.0 with 0.5 M of phosphoric acid and sterilized at 121°C for 15 min.

2.5 Fermentation.
Fermentations were carried out in 1 L capped bottles at 37°C for 72 h. Native sago starch, sago RS_3, glucose, inulin or FOS at 10% (w/v) was dispersed in 850 mL of the basal fermentation medium. Each bottle was inoculated with 10% (v/v) of fecal slurry consisted of 10^8-10^9 CFU/mL of total bacteria. Inoculated medium without a carbohydrate source served as a blank. Periodically, 20 mL of sample was withdrawn and immediately analyzed for bacterial viable count using agar plate count method. Sample was centrifuged at 2330 × *g*, 4°C for 15 min to obtain cell-free supernatant and the supernatant were analyzed for organic acids concentration and β-glucuronidase activity. The inoculation and plating were performed in an anaerobic chamber (Forma 1025 Anaerobic System, Thermo Scientific, Marietta, Ohio, USA) supplied with an atmospheric composition of 85% N_2, 10% CO_2 and 5% H_2.

2.6 Bacterial Viable Count.

One mL sample was serially diluted in 9 mL of half strength of Wilkins Chalgren Anaerobic Broth (HiMedia, Mumbai, India) in anaerobic condition. The mixture was pour plated on selective agars: nutrient agar (total aerobes); Wilkins Chalgren Anaerobic agar (total anaerobes); Bifidobacterium agar (bifidobacteria); Lactobacillus Selection agar (lactobacilli); Bacteroides Bile Esculin agar (bacteriodes); Clostridia agar (clostridia) and MacConkey agar with 0.15% Bile salts, crystal violet and NaCl (enterobacteria), incubated at 37°C for 48-72 h, and the viable count was reported as $\log_{10}$ CFU/mL. The agars were purchased either from Merck, Darmstadt, Germany, HiMedia, Mumbai, India or BD Diagnostic, Heidelberg, Germany.

2.7 Prebiotic Index

Prebiotic index of the substrates (native sago starch, sago RS, HCl-sago RS, inulin and FOS) were calculated using the equation as follows [14]:

$$\text{Prebiotic Index} = \left(\frac{\text{Bif}}{\text{Total}}\right) + \left(\frac{\text{Lac}}{\text{Total}}\right) - \left(\frac{\text{Bac}}{\text{Total}}\right) - \left(\frac{\text{Clos}}{\text{Total}}\right) \quad \text{Eq. (1)}$$

Whereby:

Bif was count of bifidobacteria at sampling time over count of bifidobacteria at 0 h of fermentation; Lac was count of lactobacilli at sampling time over count of lactobacilli at 0 h of fermentation; Bac was count of bacteroides at sampling time over count of bacteroides at 0 h of fermentation; Clos was count of clostridia at sampling time over count of clostridia at 0 h of fermentation; Total was total bacterial count (anaerobes and aerobes). Bacterial counts were expressed as colony forming unit (CFU) per mL.

2.8 pH and Organic Acids Concentration.

pH of the fermentation samples was measured by using a pH meter (Mettler Toledo, Shanghai, China). Organic acids concentrations were determined using a HPLC system [15]. Cell-free supernatant was filtered using 0.45 µm nylon membrane syringe filter (Titan 2, Sun Sri, Rockwood, Tennessee, USA) and injected onto a HPLC system (Shimadzu, Kyoto, Japan) equipped with a UV-VIS detector (Model SPD-20A, Shimadzu, Kyoto, Japan), fitted with an IC-PakTM ion-exclusion column (7.8 x 300 mm; Waters, Milford, Massachusetts, USA). The concentration of organic acids in the samples were quantified using the external standard calibration curves of lactic, acetic, succinic, propionic and butyric acids (Sigma-Aldrich, Steinheim, Germany).

2.9 β-glucuronidase activity

The cell-free supernatants were assayed for β-glucuronidase activity according to [16]. One unit of enzyme activity was equivalent to 1 µmole of ρ-nitrophenol released from ρ-nitrophenyl-β-D-glucuronide per mL at 37°C after one hour of reaction. The specific activity was expressed as unit of β-glucuronidase activity (U) per mg of protein.

2.10 Statistical analysis

The triplicate data was subjected to repeated measures analysis of variance (ANOVA), and the significance of the differences between means was determined by the Duncan test, where $p<0.05$ was considered statistically significant. The Statistical Package for Social Sciences, Version 20 (SPSS Inc., USA) was used for the analysis.

3. Results and discussion

3.1 Bacterial growths

Table 1 shows the viable counts of fecal microorganisms; total anaerobes, total aerobes, bifidobacteria, lactobacilli, bacteroides, clostridia and enterobacteria during 72 h fermentation in blank medium (without carbohydrate source) and media containing glucose, native sago starch, sago RS_3, FOS and inulin. Initially, at 0 h the fecal cultures contained an average of 7.76 ± 0.03 $\log_{10}$ CFU/ml total aerobes, 6.04 ± 0.02 $\log_{10}$ CFU/ml total aerobes, 6.54 ± 0.02 $\log_{10}$ CFU/mL bifidobacteria, 5.75 ± 0.02 $\log_{10}$ CFU/mL lactobacilli, 4.83 ± 0.01 $\log_{10}$ CFU/mL bacteroides, 7.23 ± 0.02 $\log_{10}$ CFU/mL clostridia and 4.75 ± 0.01 $\log_{10}$ CFU/mL enterobacteria.

Table 1. Bacterial population (log_{10} CFU/mL) in media containing different substrates during 72 hours of fermentation by rat fecal culture

Time (h)	Substrates					
	Blank	Glucose	Native Starch	Sago RS	FOS	Inulin
	Total Anaerobes					
0	7.77±0.11abc	7.82±0.11cd	7.77±0.17d	7.73±0.10e	7.72±0.05d	7.77±0.13e
4	7.63±0.29bc	8.72±0.20a	7.85±0.13d	8.59±0.18b	8.61±0.07b	8.47±0.16bc
8	7.78±0.24abc	8.94±0.10a	8.01±0.12bcd	8.68±0.11b	8.83±0.08a	8.58±0.10ab
12	7.96±0.17ab	8.26±0.17b	8.12±0.11abc	8.95±0.24a	8.59±0.09b	8.82±0.24a
24	8.06±0.24a	8.12±0.21bc	8.21±0.17ab	8.61±0.12b	8.29±0.18c	8.26±0.19cd
48	7.57±0.17cd	7.59±0.17de	8.28±0.15a	8.32±0.12c	7.68±0.17d	8.05±0.22de
72	7.26±0.04d	7.28±0.04e	7.94±0.03cd	8.00±0.02d	7.59±0.10d	7.84±0.11e
	Total Aerobes					
0	6.02±0.07a	6.01±0.06e	6.04±0.07b	6.06±0.02b	6.05±0.08d	6.06±0.05c
4	6.17±0.26a	7.40±0.09a	6.43±0.12ab	6.99±0.16a	7.15±0.07a	7.24±0.19a
8	6.52±0.12a	7.14±0.14ab	6.76±0.13ab	7.10±0.18a	6.63±0.14bc	6.98±0.16ab
12	6.48±0.07a	6.83±0.19bc	6.91±0.22a	7.15±0.18a	6.87±0.17ab	6.84±0.18ab
24	6.38±0.21a	6.62±0.20cd	7.03±0.17a	7.02±0.17a	6.77±0.22abc	6.91±0.12ab
48	6.14±0.49a	6.32±0.35de	6.96±0.15a	6.99±0.15a	6.54±0.20bc	6.84±0.24ab
72	5.98±0.45a	6.18±0.26e	6.90±0.27a	6.95±0.14a	6.42±0.19cd	6.69±0.27b
	Bifidobacteria					
0	6.57±0.06a	6.51±0.10c	6.55±0.05c	6.53±0.11d	6.52±0.07d	6.56±0.08d
4	6.61±0.05a	6.82±0.12bc	6.79±0.03bc	6.80±0.09c	6.81±0.04c	6.79±0.09c
8	6.70±0.11a	7.33±0.06a	6.91±0.05b	7.19±0.06b	7.24±0.11b	7.13±0.09b
12	6.70±0.11a	7.46±0.17a	7.11±0.13ab	7.82±0.14a	7.88±0.13a	7.68±0.18a
24	6.62±0.23a	7.26±0.13a	7.12±0.22ab	7.81±0.14a	7.70±0.14a	7.57±0.15a
48	6.28±0.23a	6.91±0.10b	7.39±0.25a	7.80±0.15a	7.19±0.17b	7.46±0.15a
72	6.26±0.11a	6.65±0.14bc	7.09±0.30a b	7.69±0.16a	7.17±0.18b	7.11±0.11b
	Lactobacilli					
0	5.72±0.10abc	5.76±0.10bc	5.74±0.11b	5.76±0.10c	5.74±0.11b	5.76±0.11d
4	5.89±0.26ab	6.70±0.11a	6.02±0.20ab	6.61±0.13b	6.27±0.16a	6.00±0.20cd
8	5.98±0.20a	6.89±0.12a	6.32±0.11a	6.76±0.22ab	6.37±0.12a	6.13±0.11bc
12	5.96±0.08a	6.59±0.06a	6.39±0.11a	6.96±0.17a	6.51±0.11a	6.35±0.10ab
24	5.90±0.39ab	6.10±0.24b	6.41±0.18a	6.87±0.13ab	6.46±0.28a	6.59±0.16a
48	5.50±0.06bc	5.87±0.14bc	6.22±0.15a	6.77±0.11ab	6.36±0.05a	6.51±0.09a
72	5.37±0.18c	5.70±0.14c	6.01±0.11ab	6.76±0.08ab	6.20±0.13a	5.92±0.14cd
	Bacteroides					
0	4.84±0.10c	4.83±0.09c	4.82±0.11d	4.84±0.11b	4.83±0.09c	4.82±0.11
4	4.76±0.25c	5.16±0.15b	4.93±0.12d	5.14±0.15b	5.04±0.13b	5.00±0.15c
8	5.11±0.15bc	5.36±0.21a	7.02±0.10c	6.09±0.17a	5.99±0.15a	5.29±0.16b
12	5.27±0.25ab	0d	7.23±0.05b	6.22±0.14a	4.78±0.11c	6.32±0.16a
24	5.54±0.27a	0d	7.37±0.13ab	5.00±0.17b	0d	3.73±0.16d
48	5.07±0.15bc	0d	7.42±0.11a	3.66±0.30c	0d	0e
72	4.73±0.20c	0d	7.01±0.23c	2.39±0.14d	0d	0e
	Clostridia					
0	7.24±0.13a	7.24±0.19bc	7.24±0.12c	7.20±0.13c	7.24±0.13c	7.22±0.17b
4	7.10±0.14a	8.38±0.09a	7.47±0.08ab	8.07±0.10a	8.13±0.18a	8.04±0.13a
8	7.21±0.19a	7.99±0.11a	7.50±0.09ab	7.97±0.12a	7.86±0.10ab	8.00±0.09a
12	7.20±0.21a	7.55±0.14b	7.52±0.10ab	7.88±0.10a	7.70±0.22b	7.95±0.13a
24	7.30±0.17a	6.99±0.46cd	7.55±0.11ab	7.60±0.14b	7.12±0.26c	7.22±0.20b
48	7.02±0.24a	6.68±0.33d	7.61±0.11a	7.17±0.22c	6.71±0.27d	6.88±0.38b
72	6.18±0.06b	5.66±0.12e	7.39±0.11bc	6.65±0.09d	5.74±0.20e	5.78±0.18c

Table 1. Continued

Time (h)	Blank	Glucose	Native Starch	Sago RS	FOS	Inulin
	Enterobacteria					
0	4.75±0.07c	4.75±0.10c	4.75±0.12c	4.76±0.10d	4.73±0.05c	4.75±0.08e
4	5.36±0.37bc	7.10±0.12a	5.36±0.38b	6.00±0.07b	6.80±0.10a	5.82±0.25bc
8	5.78±0.38ab	5.46±0.58b	6.11±0.23a	6.52±0.11a	5.91±0.29b	6.24±0.16a
12	6.05±0.18a	0d	6.28±0.19a	6.38±0.11ab	4.89±0.25c	6.07±0.17ab
24	5.80±0.14ab	0d	6.35±0.22a	5.99±0.11b	0d	5.60±0.24c
48	5.58±0.16ab	0d	6.43±0.11a	5.46±0.20c	0d	5.19±0.35d
72	5.33±0.15bc	0d	6.46±0.20a	5.19±0.21c	0d	5.05±0.31de

Results are expressed as means with standard deviation (N=3). Mean values in the same column followed by different lower case alphabets[abcde] within a particular substrate are significantly different ($p<0.05$) throughout the fermentation. Statistical Significance of Effect (within each group of bacteria): Effect of time, t: $p<0.05$; Effect of substrate, S: $p<0.05$; Interaction, t x S: $p<0.05$.

3.2 Selective bacterial growth as affected by different carbohydrate sources

It was observed that the growth of bifidobacteria and lactobacilli were sustained until 48 h of fermentation whereas bacteroides, clostridia and enterobacteria had started to show a significant decrease ($p<0.05$) at 24 h, 12 h and 24 h of fermentation, respectively. Sago RS_3 media exhibited higher growths of the beneficial bacteria than FOS and inulin suggesting that sago RS_3 is a potential prebiotic. Media containing sago RS_3 could sustain the growth of beneficial bacteria for 48 h of fermentation with a concomitant impediment in the counts of detrimental bacteria. In a previous *in vivo* research, an increase in the growth of bifidobacteria and lactobacilli was observed in the rat caecal content with a concomitant decrease in the counts of enterobacteria after 4 weeks of consuming retrograded high amylose corn starch [17]. However, neither *in vitro* nor *in vivo* studies had investigated the selective fermentation of sago RS_3.

3.3 Prebiotic index (PI)

PI is a quantitative approach used to magnitude the prebiotic effect of a substrate. Positive PI value indicates that the substrate exerts an increase in the counts of beneficial bacteria (bifidobacteria and/or lactobacilli) whereas negative value of PI indicates an increase in the counts of detrimental bacteria, which were bacteroides and/or clostridia [14]. Generally, all the substrates exhibited prebiotic characteristics with positive PIs were achieved by the substrates after 12-48 h of fermentation (Figure 1). The highest PI ($p<0.05$) was displayed by sago RS_3 (+12.19), followed by FOS (+9.45) and inulin (+6.82). PI of sago RS_3 at 48 h (+7.00) was not significantly different from that of inulin. Thus, this suggested that sago RS_3 could pronounce prebiotic effect similar to the commercial prebiotics.

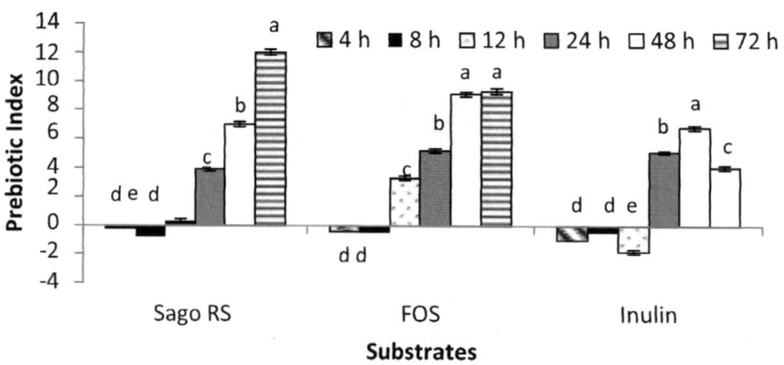

Figure 1. Prebiotic index of sago RS, HCl-sago RS, FOS and inulin. Error bars represent standard deviations of N=3. Bars bearing different alphabets within a particular substrate are significantly different ($p<0.05$). Value with * is significantly the highest ($p<0.05$) index. Statistical Significance of Effect: Effect of time, t: $p<0.05$; Effect of substrates, S: $p<0.05$; Interaction, t x S: $p<0.05$.

3.4 pH and organic acids production

Generally, all media with the addition of carbohydrate sources could show a significant decrease ($p<0.05$) in the pH value from the initial pH of ~6.52 to pH 3.38 (Data not shown). Reducing the fermentation pH is regarded as the mechanism of action of dietary carbohydrate attributed to the production of organic acids [18].

Overall, all the media display significant ($p<0.05$) increase in the production of lactic and short chain fatty acids (acetic, propionic and butyric acids) (Table 2).

Table 2. Molar concentration of organic acids (mM) in media containing different substrates during 72 hours of fermentation by rat fecal culture

Time (h)	Blank	Glucose	Native Starch	Sago RS	FOS	Inulin
	Lactic acid					
0	0c	0g	0d	0f	0g	0g
4	0c	18.56±0.10f	0.58±0.03b	12.80±0.24e	16.24±0.26f	10.99±0.13f
8	1.95±0.10a	37.15±0.23e	1.96±0.03a	20.82±0.04d	30.20±0.64e	14.02±0.60e
12	1.42±0.10b	59.73±0.5d	0.28±0.01c	27.48±0.05a	38.66±0.23d	20.74±0.55d
24	0c	68.20±1.03c	0.00d	25.65±0.52b	56.20±0.44c	34.37±0.67c
48	0c	75.60±0.45b	0.00d	24.80±1.01b	64.73±0.09b	50.69±1.69b
72	0c	79.08±1.37a	0.00d	23.27±0.64c	75.42±1.15a	55.34±0.73a
	Acetic Acid					
0	0g	0f	0f	0e	0g	0f
4	0.79±0.04f	4.90±0.05e	1.25±0.02e	4.97±0.17b	3.62±0.08f	1.24±0.04e
8	1.38±0.03e	5.40±0.07d	8.08±0.05d#	5.17±0.10ab	4.47±0.17e	3.57±0.15d
12	3.80±0.06d	5.96±0.03c	8.10±0.03d#	5.42±0.05a	5.91±0.06d	4.89±0.07c
24	4.17±0.13c	6.77±0.15b	11.84±0.14c*#	4.50±0.13c	6.86±0.09c	6.01±0.27b
48	4.84±0.22b	6.74±0.15b	16.90±0.38b*#	3.38±0.40d	7.94±0.08b	6.04±0.12b
72	5.42±0.19a	7.02±0.17a	17.46±0.43a*#	3.08±0.08d	8.97±0.07a	6.54±0.18a
	Propionic Acid					
0	0e	0g	0g	0f	0g	0n
4	0e	1.28±0.04f	1.77±0.06f	2.37±0.09e	1.09±0.04f	0.66±0.04f
8	1.23±0.04d	4.28±0.07e	3.89±0.11e	6.38±0.26d	2.14±0.02e	1.62±0.10e
12	2.00±0.06c	4.54±0.05d	5.01±0.06d	9.98±0.47c	3.85±0.09d	3.11±0.15d
24	2.48±0.09b	8.89±0.16c	9.17±0.25c	18.87±0.80b*	7.07±0.08c	10.85±0.43c
48	2.47±0.07b	12.15±0.12b	9.48±0.06b	20.16±0.58a*	13.73±0.52b	16.39±0.39b
72	3.39±0.20a	13.93±0.24a	11.55±0.04a	20.84±0.26a*	16.60±0.22a	26.74±0.10a
	Butyric Acid					
0	0g	0g	0g	0f	0g	0f
4	0.12±0.01f	0.03±0.00f	0.04±0.00f	0.02±0.00f	0.74±0.01f	0.11±0.00e
8	0.17±0.01e	0.05±0.00e	0.20±0.01e	0.32±0.01e	0.87±0.00e	0.20±0.02d
12	0.29±0.01d	0.34±0.01d	0.23±0.01d	1.98±0.08d	1.66±0.01d	0.27±0.01d
24	0.42±0.02c	0.97±0.02c	0.80±0.01c	2.92±0.04c*#	1.70±0.01c	1.56±0.09c
48	0.78±0.00b	1.50±0.01b	1.83±0.01b	4.08±0.02b*#	1.86±0.02b	1.98±0.02b
72	0.82±0.00a	1.61±0.02a	2.21±0.01a*	4.21±0.12a*#	1.90±0.03a	2.47±0.06a

Results are expressed as means with standard deviation (N=3). Mean values in the same column followed by different lower case alphabets within a particular substrate are significantly different ($p<0.05$) throughout the fermentation. Statistical Significance of Effect: Effect of time, t: $p<0.05$; Effect of substrate, S: $p<0.05$; Interaction, t x S: $p<0.05$. Within a particular type of short chain fatty acid, *values are significantly higher ($p<0.05$) than the highest concentration of acid in media containing FOS at 72 h while #values are significantly higher ($p<0.05$) than the highest concentration of acid in media containing inulin at 72 h.

Acetic, propionic and butyric acids are the predominant short chain fatty acids generated in the human colon, accounting for 85-95% of the total short chain fatty acids produced [19]. Accumulation of lactic acid was seen in media containing glucose, FOS and inulin throughout the fermentation period and reached the maximal level at 72 h of fermentation with the highest concentration of 79.08

mM, 75.42 mM and 55.34 mM, respectively. The concentration of acetic acid in media containing sago RS_3 increased significantly ($p<0.05$) before a depletion occurred at the 24 h of fermentation and the concentration of propionic acid in media containing sago RS_3 (18.87-20.84 mM) was significantly higher ($p<0.05$) after 24 h than that of by FOS (16.60 mM) at 72 h of fermentation. The production of butyric acid increased continuously throughout the fermentation period with the highest concentrations of 4.21 mM (sago RS_3), 2.47 mM (inulin), 2.21 mM (native starch), 1.90 mM (FOS) and 1.61 mM (glucose) at 72 h of fermentation. The highest concentration of butyric acid production by sago RS_3 was significantly higher ($p<0.05$) than that of by the commercial prebiotics, FOS and inulin.

3.5 β-glucuronidase activity

Media containing glucose, sago RS_3, FOS and inulin showed significant decrease ($p<0.05$) in the activity of β-glucuronidase throughout the fermentation period whereas blank medium and medium with native sago starch exhibited a significant increase ($p<0.05$) until 48 h of fermentation concomitant with the growth of bacteroides, clostridia and enterobacteria as they are the producers of β-glucuronidase (Figure 2).

Patients diagnosed with colon cancer were found to have a higher activity of fecal β-glucuronidase than that of healthy people [20]. Thus, lowering β-glucuronidase activity would be desirable as this enzyme plays a crucial role in promoting carcinogenesis.

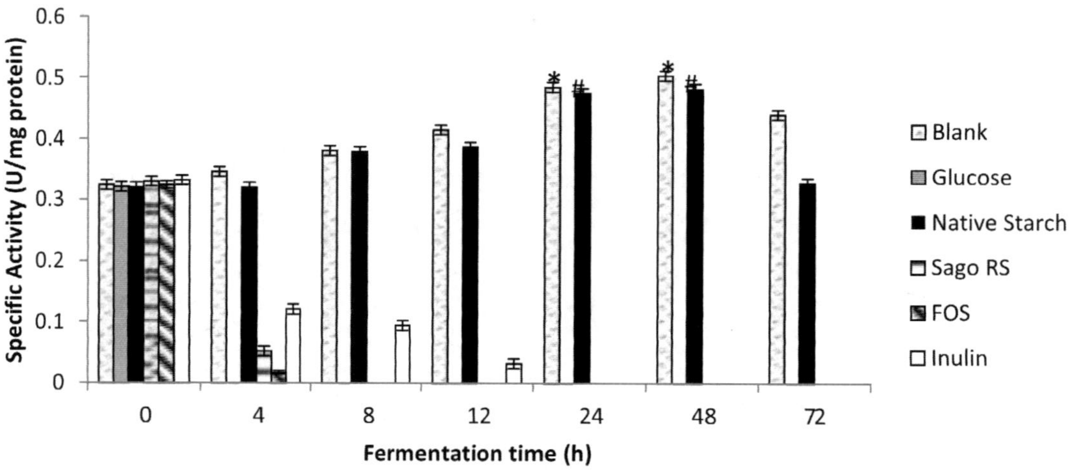

Figure 2. Activity of β-glucuronidase during fermentation of different substrates by rat fecal cultures. Error bars represent standard deviations of N=3. Bars bearing same symbols (*, #) within same substrate are not significantly different ($p>0.05$). Statistical Significance of Effect: Effect of time, t: $p<0.05$; Effect of substrates, S: $p<0.05$; Interaction, t x S: $p<0.05$.

4. Conclusions

Sago RS_3 exhibited comparable prebiotic effects with commercial prebiotics FOS and inulin, whereby sago RS_3 stimulated the growth of beneficial bacteria (lactobacilli and bifidobacteria) at the expense of detrimental bacteria (bacteroides, clostridia and enterobacteria). Sago RS_3 displayed the highest prebiotic index and fermentation of sago RS_3 produced the highest concentration of butyric acid while reduced β-glucuronidase activity.

References

[1] Sekirov, I, Russell, S L, Antunes, L C and Finlay, B B 2010 Gut microbiota in health and disease *Physiol Rev* **90**(3) 859-904.

[2] Zhu, Q, Jin, Z, Wu, W, Gao, R, Guo, B, Gao, Z, Yang, Y, and Qin, H 2014 Analysis of the intestinal lumen microbiota in an animal model of colorectal cancer *PLoS One* **9**(3)e90849.

[3] Gibson, G R and Roberfroid, M B 1995 Dietary modulation of the human colonic microbiota: introducing the concept of prebiotics *J Nutr* **125**(6) 1401-12.

[4] Gibson, G R, Scott, K P, Rastall, R A, Tuohy, K M, Hotchkiss, A, Dubert-Ferrandon, A, Gareau, M, Murphy, E F, Saulnier, D, Loh, G, Macfarlene, S, Delzenne, N, Ringel, Y, Kozianowski, G, Dickman, R, Lenoir-Wijnkoop, I, Walker, C, and Buddington, R 2010 Dietary prebiotics: current status and new definition *Food Sc Tech Bull:Funct Food* **7**(1) 1-19.

[5] Gibson, G. R, Probert, H M, Loo, J V, Rastall, R A and Roberfroid, M B 2004 Dietary modulation of the human colonic microbiota: updating the concept of prebiotics *Nutr Res Rev* **17**(2) 259-75.

[6] Newswire, P R 2015 Global prebiotics ingredients (oligosaccharides, inulin, polydextrose and others) Market 2015-2020 - China to dominate global prebiotic ingredients Accessed on 15 August 2015 from http://www.prnewswire.com/news-releases/global-prebiotics-ingredients-oligosaccharides-inulin-polydextrose-others-market-2015-2020-china-to-dominate-global-prebiotic-ingredients-300119678.html.

[7] Davani-Davari D, Negahdaripour M, Karimzadeh I, Seifan M, Mohkam M, Masoumi S J, Berenjian A and Ghasemi Y 2019 Prebiotics: Definition, Types, Sources, Mechanisms and Clinical Applications Foods **8**(92) 1-27.

[8] Ashwar, B A, Gani, A, Shah, A, Wani, I A and Masoodi, F A 2015 Preparation, health benefits, and applications of resistant starch - a review *Starch/Stärke* **67** 1-15.

[9] Zi-Ni, T, Rosma, A, Napisah, H, Karim, A. A, and Liong, M T 2015a Characteristics of *Metroxylon sagu* resistant starch type III as prebiotic substance *J Food Sc* **80**(4) H875-82.

[10] Baghurst, P A, Baghurst, K L and Record, S J 2001 *Handbook of dietary fibre in human health,* ed G A Spiller (Florida: CRC Press) pp 583-91.

[11] Al-Tamimi, M A H M, Palframan, R J, Cooper, J M, Gibson, G R and Rastall, R A 2006. *In vitro* fermentation of sugar beet arabinan and arabino-oligosaccharides by the human gut microflora *J Appl Microbiol* **100**(2) 407-14.

[12] Zi-Ni, T, Rosma, A, Karim, A A and Liong, M. T. (2015b). Functional properties of resistant starch type-III from *Metroxylon sagu* as affected by processing conditions *Pertanika J Trop Agr Sc* **38**(3) 399-412.

[13] Lesmes, U, Beards, E J, Gibson, G R, Tuohy, K M and Shimoni, E 2008 Effects of resistant starch type III polymorphs on human colon microbiota and short chain fatty acids in human gut models *J Agr Food Chem* **56**(13) 5415-21.

[14] Palframan, R, Gibson, G R and Rastall, R A 2003 Development of a quantitative tool for the comparison of the prebiotic effect of dietary oligosaccharides *Lett Appl Microbiol* **37**(4) 281-4.

[15] Rycroft, C E, Jones, M R, Gibson, G R and Rastall, R A 2001 A comparative in vitro evaluation of the fermentation properties of prebiotic oligosaccharides *J Appl Microbiol* **91**(5) 878-87.

[16] Shen, R L, Dang, X Y, Dong, J L, and Hu, X Z 2012 Effects of oat β-glucan and barley β-glucan in fecal characteristic, intestinal microflora, and intestinal bacterial metabolites in rats *J Agr Food Chem* **60**(45) 11301-8.

[17] Silvi, S, Rumney, C J, Cresci, A and Rowland, I R 1999 Resistant starch modifies gut microflora and microbial metabolism in human flora-associated rats inoculated with faeces from Italian and UK donors *J Appl Microbiol* **86**(3) 521-30.

[18] Walker, A W, Ince, J, Duncan, S H, Webster, L M, Holtrop, G, Ze, X, Brown, D, Stares, M D, Scott, P, Bergerat, A, Loius, P, McIntosh, F, Johnstone, A M, Lobley, G E, Parkhill, J and Flint, H J 2011 Dominant and diet-responsive groups of bacteria within the human colonic microbiota *ISME J* **5**(2) 220-30.

[19] Cummings, J H and Macfarlane, G T 1991 The control and consequences of bacterial fermentation in the human colon *J Appl Bacteriol* **70**(6) 443-59.

[20] Kim, D H and Jin, Y H 2001 Intestinal bacteria β-glucuronidase activity of patients with colon cancer *Arch Pharmacal Res* **24**(6) 564-7.

Acknowledgements
This research was funded Universiti Sains Malaysia RU Grant 1001/PTEKIND/811209 and Fellowship Scheme. We are thankful to DPO Malaysia Pvt. Ltd. for providing FOS and inulin.

IOP Conf. Series: Materials Science and Engineering **716** (2020) 012003 doi:10.1088/1757-899X/716/1/012003

Screening of Medium Constituents for the Cultivation of *Scenedesmus dimorphus* UTEX 1237 using 2^k Factorial Design Approach

B Ravindran[1], M A Kassim[2], and M S Mohamed[1,3]

[1] Department of Bioprocess Technology, Faculty of Biotechnology and Biomolecular Sciences, Universiti Putra Malaysia, 43400 UPM Serdang, Malaysia.

[2] Bioprocess Technology, School of Industrial Technology, Universiti Sains Malaysia, 11800 Gelugor, Pulau Pinang, Malaysia.
[3] Bioprocessing and Biomanufacturing Research Centre, Faculty of Biotechnology and Biomolecular Sciences, Universiti Putra Malaysia, 43400 UPM Serdang, Malaysia.

Email: m_shamzi@upm.edu.my

Abstract. Freshwater microalga, *Scenedesmus dimorphus* is a beneficial natural resource for producing biofuel and novel bioproducts. *S. dimorphus* growth greatly depends on physical culture conditions as well as medium constituents. A study was conducted to select the suitable medium components that could promote high biomass production in *S. dimorphus* UTEX 1237. Screening of growth medium constituents was performed using the two-level factorial design (2^k FD) method. *S. dimorphus* was cultivated in TAPS aquaculture medium employing two types of carbon source (sodium bicarbonate and acetic acid) and three types of nitrogen sources (ammonium chloride, sodium nitrate, and urea). Consequently, medium components that significantly influenced biomass production comprised of acetic acid, ammonium chloride, and urea. Two weeks of *S. dimorphus* cultivation in shake flask system consisting of 1.5 g/L of acetic acid, 10 g/L of ammonium chloride and 0.4 g/L of urea was found to affect a specific growth rate of 0.0049 h^{-1} and biomass productivity of 0.05 g/L/day, thus affecting 66.7% improvement over normal TAP formulation.

1. Introduction

The immense diversity of microalgae has traditionally drawn the attention of scientific communities into manipulating them for human consumptions. Given the increasing demand leading to the continuous spike in petroleum pricing and the alarming impact of global warming due to the burning of fossil fuels, numerous reports were published extolling the advantages of shifting to microalgal biofuel on the basis of low energy requirement for production, non-toxicity, biodegradability and low carbon dioxide emissions into the atmosphere [1]. Microalgae biomass is composed of major macromolecules such as lipids, carbohydrates, and proteins which can be selectively converted into liquid biodiesel, bioethanol, biomethane, biohydrogen, and bio-oil by subjecting them to different biological conditions [2]. Microalgae proved a promissory source of biofuel since it results in higher yields with respect to other lignocellulosic biomass, being capable of producing 100 times more energy per hectare compared to the well-known terrestrial crops, i.e., corn, soybean, rapeseed, oil palm and jatropha [3].

Some of the biotechnologically relevant microalgae species are *Chlorella, Spirulina, Dunaliella, Haematococcus, Isochrysis,* and *Scenedesmus* [4]. *Scenedesmus* spp. are appropriate to be used for laboratory-scale studies and mass intensive cultivation since they have a high tolerance to differences in light intensity, pH, temperature, and high concentration of carbon dioxide [5]. The utilization of *Scenedesmus* biomass has advances beyond the field of biofuel as they are now being incorporated into plastic formulation or used as cell factory to produce PLA or PHA bioplastics [6], biogenic synthesis of silver nanoparticles [7], source of biocement or bioconcrete [8], and natural bioactive compounds (NBC) for treatment and prevention of human diseases [9].

Large scale microalgae cultivation for biofuel or any of the specialized chemicals might not be economically viable as much of the total costs are concentrated in the production of microalgae biomass (50%) whereas another 20%-30% is allocated to the biomass recovery process (harvesting and drying) [10]. Low biomass concentration will imply greater labor and energy requirement for dewatering effort. The ability of microalgae to increase their biomass density greatly depends on the cultivation conditions and growth medium. Natural medium such as soil medium is quite cheap to produce, but it does not necessarily contain all the nutrients required for the growth of microalgae. Thus, an artificial formulation is introduced whereby the nutrient composition and concentration are manipulated according to the species of interest. Improvement of medium composition can be achieved through proper screening and optimization exercise.

Based on the number of factors involved, sometimes the total experimental runs to be performed can be quite daunting if the traditional varying one-factor-at-a-time (OFAT) method is practiced. Therefore, statistical-based experimental design such as the two-level factorial design (2^k FD) or Plackett-Burman is usually employed to simplify the process. These techniques are efficient for the screening of significant factors from multiple dependent variables and to determine the existence of potential interactions between them, a boon which traditional OFAT experiments were unable to demonstrate precisely [11]. In this study, the formulation of enhanced TAP growth medium specifically for freshwater microalga, *Scenedesmus dimorphus* UTEX 1237 was focused on different types and concentrations of carbon and nitrogen sources. Screening of significant medium constituents affecting specific growth rate and final *S. dimorphus* biomass concentration is reported using two-level factorial design.

2. Materials and Methods

2.1. Microalga Species

Scenedesmus dimorphus UTEX 1237 was originally sourced from the UTEX Culture Collection, University of Texas, United States. The stock culture was maintained and sub-cultured monthly in TAP medium slant agar and kept at the Bioprocessing and Biomanufacturing Research Centre (BBRC), Universiti Putra Malaysia under 4°C and exposed to light and dark photoperiod of 16h: 8h.

2.2. Inoculum and Seed Culture Preparation

For uses as inoculum, a loop-full of microalga colony from slant agar was transferred into 50 mL Falcon tube containing 10 mL of sterilized TAP medium. The inoculum was incubated at 25 °C in an orbital incubator shaker (Model IS-971 Lab Companion), agitated at 120 rpm for 14 days. Illumination was provided by T5 fluorescent lamps with light and dark photoperiod of 16h: 8h as well. For subsequent seed culture preparation, 10 mL of microalgae culture was then transferred into 250 mL Erlenmeyer flasks containing 90 mL of TAP medium and incubated for another 14 days.

2.3. Medium for Microalga Growth

Tris-Acetate-Phosphate (TAP) medium [12] was used as a basal medium to cultivate *S. dimorphus*. Two types of carbon sources (acetic acid and sodium bicarbonate) and three types of nitrogen sources

IOP Conf. Series: Materials Science and Engineering **716** (2020) 012003 doi:10.1088/1757-899X/716/1/012003

(sodium nitrate, urea, and ammonium chloride) were tested. All microbiological-grade carbon and nitrogen sources used were stable when sterilized at 121°C, 15 psi for 15 min. Some of the components were autoclaved separately to prevent any Maillard reaction. All cultivations were performed in 100 mL medium with 10% (v/v) inoculum from seed culture prepared earlier. The cultures were grown in an orbital incubator with all conditions maintained and incubation period set for 21 days.

2.4. Experimental Design
Carbon and nitrogen sources in the original TAP medium comprised of acetic acid and ammonium chloride. This study assessed the effectiveness of other alternatives to medium formulation. Two-level factorial design (2^k FD) was used to screen for the most significant factors influencing the *S. dimorphus* biomass yield. As illustrated in Table 1, five variables factors were chosen, which were sodium bicarbonate (X_1), acetic acid (X_2), ammonium chloride (X_3), sodium nitrate (X_4) and urea (X_5), with each factor having two levels of concentration: -1 (low level) and +1 (high level). Each component's concentration at these two levels was taken from the most minimum and maximum values utilized by other investigators in literature for *Scenedesmus* species. The main response variable was biomass concentration (g/L). Design-Expert (Version 7.0.0, Stat-Ease Inc, Minneapolis, USA) software was used for designing the experiment as well as analysis of results. Variables which imparted a major effect on elevating S. dimorphus biomass concentration were identified based on confidence level above 95% ($P < 0.05$). The significant factors and possible interactions identified from half-normal plot analysis were chosen for generation of the first-order model for the response as per Equation (1) following evaluation of effects and interaction by analysis of variance (ANOVA):

$$Y = B_0 + B_1X_1 + B_2X_2 + B_3X_3 + B_4X_4 + B_5X_5 \quad (1)$$

Where Y is response; B_0 is modal constant, and B_i is the linear coefficient. The variable, X_i is the non-coded independent variables. The coefficient of determination, $R^{2,}$ and adjusted R^2 was used for evaluation of the fitness of the model.

Table 1. Experimental range and levels of independent variables

Variables (g/L)	Symbol	Actual levels of coded factors	
		-1	+1
Sodium bicarbonate	*A*	0.40	10.0
Acetic acid	*B*	0.50	1.50
Ammonium chloride	*C*	10.0	16.0
Sodium nitrate	*D*	0.20	5.00
Urea	*E*	0.40	9.00

2.5. Microalga Growth Analysis
Biomass measurement of *S. dimorphus* UTEX 1237 was obtained through the gravimetric method. 2 mL of culture samples taken at every two-day interval were transferred into pre-weight clean centrifuge tubes and centrifuged at 10 000 rpm for 10 minutes (Eppendorf 5415 D mini centrifuge, Germany). Upon supernatant removal, individual cell pellet together with centrifuge tube would be allowed to dry at 95°C in the oven for several days and weighed until constant dried cell weight was achieved. The biomass measurement was calculated in terms of g/L.

2.6. Growth Kinetics Calculation
The time-specific measurements of *S. dimorphus* UTEX 1237 dried cell weight were fitted to the Baranyi-Roberts microbial propagation model [13]. The dynamic model calculation of growth kinetic parameters was made possible via online curve-fitting DMFit software (www.combase.cc), in which

the lag phase duration (λ, h) and maximum specific growth rate during exponential phase (μ, h^{-1}) were estimated.

3. Results and Discussion

3.1. Screening of Medium Constituents Affecting S. dimorphus Biomass Concentration

Screening is a crucial step in eliminating components that do no influence the response. Thus, this procedure ensures that the medium composition provided is utmostly favorable for the growth of microalgae and to accumulate the product of interest in desired amounts. In this study, two-level factorial (2^k FD) is preferred over Plackett-Burman (PB) design as 2^k FD is a more comprehensive tool in investigating each complete trial or replication of the experiment with all possible combination of factors' levels and the interaction between each factor. Screening by PB design is more economical to identify a few main factors affecting the outcome, only when the interaction between factors are not significant. Hence, if there are significant interactions, PB design could provide misleading results.

Besides acetic acid and ammonium chloride, three more commonly employed medium components for aquaculture purposes were identified from previous research in the literature. The five-variable experimental design using 2^k FD had created a total of 32 cultivation runs (N = 2^5), as shown in Table 2.

Table 2. Two-level factorial design showing actual levels and responses

Std order run	*A*	*B*	*C*	*D*	*E*	Dried Cell Weight (g/L)	
						Actual	Predicted
1	0.40	0.5	10.0	0.2	0.4	0.554	0.57
2	10.0	0.5	10.0	0.2	0.4	0.585	0.57
3	0.40	1.5	10.0	0.2	0.4	0.632	0.66
4	10.0	1.5	10.0	0.2	0.4	0.706	0.68
5	0.40	0.5	16.0	0.2	0.4	0.319	0.38
6	10.0	0.5	16.0	0.2	0.4	0.342	0.38
7	0.40	1.5	16.0	0.2	0.4	0.530	0.46
8	10.0	1.5	16.0	0.2	0.4	0.522	0.46
9	0.40	0.5	10.0	5.0	0.4	0.617	0.57
10	10.0	0.5	10.0	5.0	0.4	0.595	0,57
11	0.40	1.5	10.0	5.0	0.4	0.595	0.66
12	10.0	1.5	10.0	5.0	0.4	0.638	0.66
13	0.40	0.5	16.0	5.0	0.4	0.339	0.38
14	10.0	0.5	16.0	5.0	0.4	0.350	0.38
15	0.40	1.5	16.0	5.0	0.4	0.520	0.46
16	10.0	1.5	16.0	5.0	9.0	0.459	0.46
17	0.40	0.5	10.0	0.2	9.0	0.402	0.34
18	10.0	0.5	10.0	0.2	9.0	0.373	0.34
19	0.40	1.5	10.0	0.2	9.0	0.499	0.42
20	10.0	1.5	10.0	0.2	9.0	0.355	0.42
21	0.40	0.5	16.0	0.2	9.0	0.309	0.29
22	10.0	0.5	16.0	0.2	9.0	0.328	0.29
23	0.40	1.5	16.0	0.2	9.0	0.359	0.37
24	10.0	1.5	16.0	0.2	9.0	0.292	0.37
25	0.40	0.5	10.0	5.0	9.0	0.398	0.34
26	10.0	0.5	10.0	5.0	9.0	0.295	0.34
27	0.40	1.5	10.0	5.0	9.0	0.399	0.42
28	10.0	1.5	10.0	5.0	9.0	0.312	0.42

29	0.40	0.5	16.0	5.0	9.0	0.279	0.29
30	10.0	0.5	16.0	5.0	9.0	0.239	0.29
31	0.40	1.5	16.0	5.0	9.0	0.502	0.38
32	10.0	1.5	16.0	5.0	9.0	0.310	0.37

Choosing which of the five variables deemed significant to *S. dimorphus* growth requires observation of the half-normal plot and ANOVA table. Upon fitting to a multiple regression model as in Equation (1), p-value generated by ANOVA table will determine whether the model, as a whole, is significant. A backward elimination procedure was carried out during variable selection to reduce the number of unnecessary model terms (usually by removing variables or interaction terms one at a time, beginning with the term having the highest p-value). Continuous subtraction of these terms would consequently update the overall model's p-value. Elimination was stopped when only terms with p-values below the acceptable threshold of 0.05 were left in the model, resulting in a reduced model as in Table 3.

Table 3. ANOVA table of the reduced factorial model following backward elimination

Source	Sum of Squares	Degrees of freedom	Mean Square	F-value	P > F
Model	0.46	6	0.077	30.44	<0.0001
B- Acetic acid	0.053	1	0.053	21.01	0.0001
C- Ammonium chloride	0.12	1	0.12	47.23	<0.0001
E-Urea	0.22	1	0.22	86.64	<0.0001
CE	0.040	1	0.040	15.62	0.0023
Residual	0.094	27	3.493E-003		
Cor Total	0.53	31			

The resulting relationship of all significant factors toward *S. dimorphus* growth was represented by a linear regression Equation (2). The fitness of the model, as expressed by R^2, is 0.8211, indicating that 82.11% of the variability in the response can be explained by this linear model. In addition, the goodness of fit of the regression equation is also tested by looking at adjusted R-squared. R^2_{Adj} is 0.7946, and the predicted R-squared is 0.7487. The R^2_{Adj} which is 0.7946, indicates a still high degree of agreement between the observed and predicted values for the response studied, suggesting that the proposed model equation can provide satisfactory results.

$$\text{Biomass concentration (g/L)} = 0.8770 + 0.0816B - 0.0332C - 0.0548E + 2.729 \times 10^{-3}\,CE \qquad (2)$$

The half-normal plot determines the significant factors affecting the response. Graphically, these factors are represented by the points that deviate far to the right-side from the linear line in the plot (Figure 1). Urea (*E* factor) is the furthest, indicating the strongest signal, then followed by ammonium chloride (*C* factor), acetic acid (*B* factor) and interaction between ammonium chloride and urea (*CE* factor). This interpretation concurs with F-values ranking in the ANOVA table. Contribution-wise, it was calculated by Design Expert that the effect of urea was 41.72%, ammonium chloride was 22.74%, acetic acid was 10.12%, and the *CE* interaction term was 7.52% towards the response.

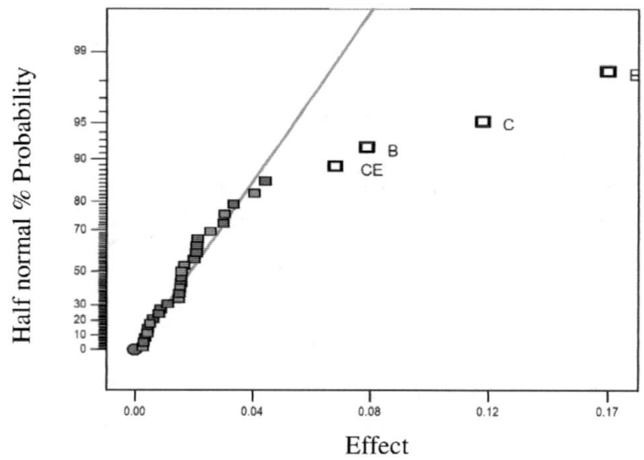

Figure 1. Half-normal plot depicting the selection of the main effect of acetic acid (*B*), ammonium chloride (*C*) and urea (*E*).

3.2. Response Analysis of S. dimorphus UTEX 1237 Biomass Concentration

The effect of each major factor, as well as any interaction among factors towards influencing S. dimorphus biomass propagation, were statistically investigated. As mentioned previously, one carbon source; acetic acid (*B* factor) and two nitrogen sources; ammonium chloride (*C* factor) and urea (*E* factor) have a significant statistical effect whereas sodium bicarbonate (*A* factor) and sodium nitrate (*D* factor) supplemented at given concentration range did not noticeably affect the cell growth. Moreover, the positive and negative effects of these three important variables are presented in Figure 2.

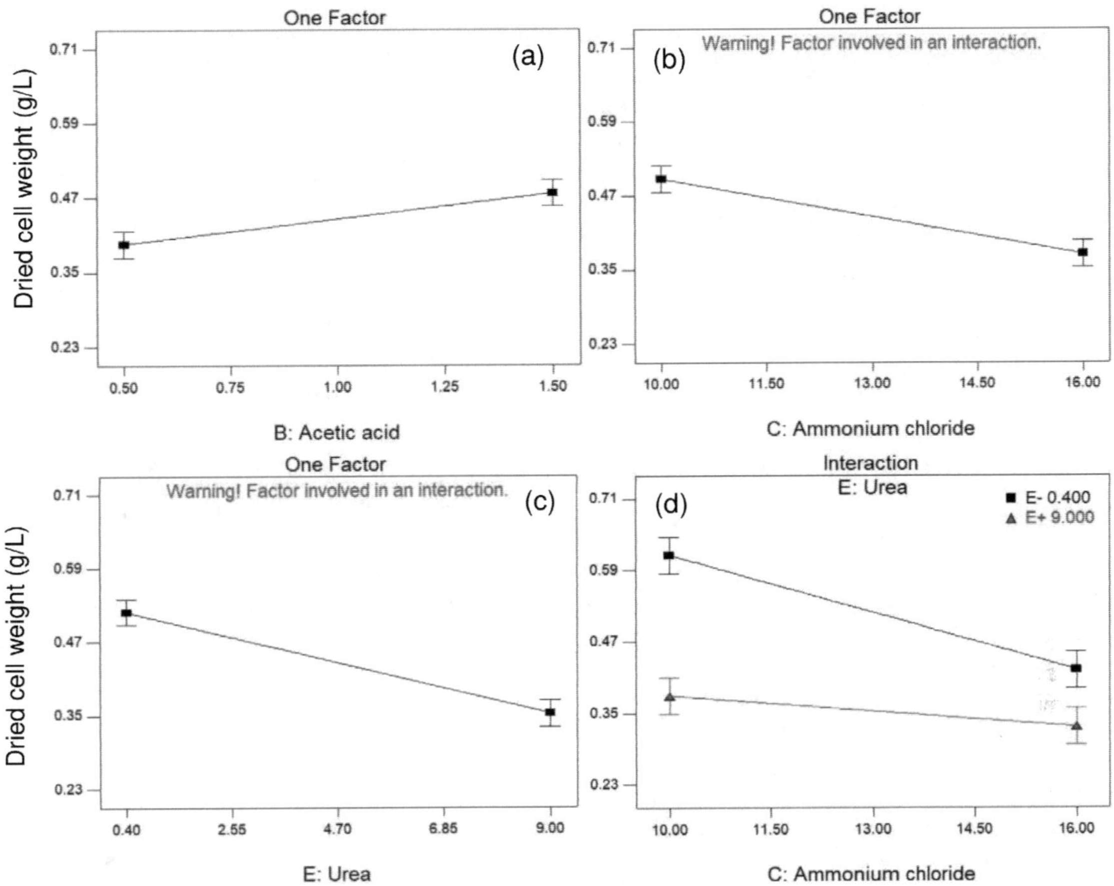

Figure 2. The plot of significant factors and interaction at 95% confident interval

As for major factors, acetic acid in Figure 2(a) posed a positive effect with an increased in concentration level, improving biomass concentration from 0.395 g/L to 0.477 g/L. Nonetheless, both ammonium chloride and urea imparted negative. Interestingly, significant mutual effect occurred between both nitrogen sources when *S. dimorphus* biomass was found higher at low urea concentration of 0.4 g/L coupled with low level of ammonium chloride of 10 g/L compared to a high point of 16 g/L, affecting substantial dried cell weight of 0.615 g/L against individual effect of low level ammonium chloride (0.497 g/L) and urea (0.512 g/L). The biomass yield reduced with an increase of both medium constituents. Thus, the optimal TAP medium composition from 2^k FD screening was registered by standard order run no. 4 in Table 2 which comprised of 1.5 g/L of acetic acid, 10 g/L of ammonium chloride and 0.4 g/L of urea giving a final yield of biomass at 0.706 g/L.

3.3. Comparison of Microalga Growth between 2^k FD and Original TAP Formulation

The time course data of *S. dimorphus* UTEX 1237 growth for cultivation that had employed formulation based on the outcome of 2^k FD screening together with a culture grown in normal TAP medium is shown in Figure 3.

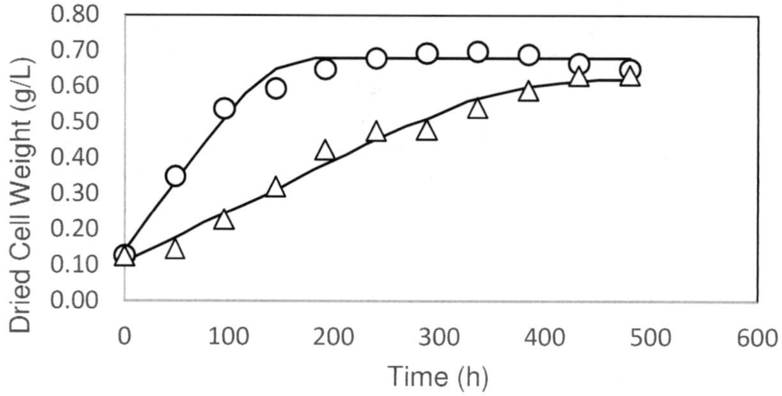

Figure 3. Growth profile of *S. dimorphus* UTEX 1237 on the modified medium from 2^k FD screening (circle) and original TAP (triangle). The solid line represents growth as simulated by Baranyi-Robert kinetic model

No apparent lag phase was observed in both cultures, indicating rapid adapting of *S. dimorphus* cells from seed to production medium. There are distinct regions defining the exponential to stationary growth phase around Day 6 in culture tested with the newly modified medium. The original formulation, however, shows a much gradual exponential growth and seemed to decelerate around Day 16. From kinetic model calculation, this translates to improved maximum specific growth, μ_{max} of 0.0049 h^{-1} for the modified medium against 0.00141 h^{-1} in normal TAP. Improvement of biomass saw the highest density was achieved on Day 14 at 0.706 g/L (productivity at 0.050 g/L/day) whereas in the original TAP it was 0.633 g/L gained at the end of 21 days run (productivity at 0.030 g/L/day).

Acetic acid content was 1.05 g/L in the original in TAP medium. By adding 0.45 g/L more to the medium, it significantly improved the mixotrophic growth of *S. dimorphus*. However, any more addition of substrate concentration should be done carefully, as the nature of this organic acid or bicarbonate, by algal consumption would release carbon dioxide in the water to form carbonic acid that can decrease the culture pH [14]. *S. dimorphus* affinity towards acetic acid is encouraging since this substrate can be synthesized inexpensively and in great quantities from petroleum resources [15].

The preference towards ammonium ion instead of nitrate might due to the strains having low nitrate reductase activity for reducing nitrate via nitrite to ammonium. The highest contribution to response is shown by urea at 0.4 g/L. According to Sakamoto *et al.* [16], urea is a better nitrogen source for the cultivation of microalgae than nitrates and ammonia due to less energy is required for the assimilation process and having the ability to pass through the plasma membrane of microalgae cells. Most fast growing microalgae preferred ammonia as their nitrogen source, while urea, a reduced form of nitrogen source can also be readily hydrolyzed into ammonia, lending to the synergistic effect toward the cell growth by having two nitrogenous-rich compounds in the medium.

Conclusion

Two-level factorial design on five TAP medium constituents has shown that *S. dimorphus* UTEX 1237 biomass concentration was significantly affected by acetic acid, urea and ammonium chloride. Additionally, a significant interaction existed between the two nitrogen sources. The proposed medium composition for enhancing yield was as follow; 1.5 g/L of acetic acid, 10 g/L of ammonium chloride and 0.4 g/L of urea, which achieved final biomass productivity of 0.050 g/L/day, translating to 66.7% improvement over normal TAP formulation.

References

[1] de Morais M G, de Freitas B C B, Moraes L, Pereira A M, Costa J A V 2019 Liquid biofuels from microalgae: recent trends *Advanced Bioprocessing for Alternative Fuels, Biobased Chemicals, and Bioproducts* ed M Hosseini (Cambridge: Woodhead Publishing) chapter 18 pp 351–72

[2] Costa J A V and de Morais M G 2011 The role of biochemical engineering in the production of biofuels from microalgae *Bioresour. Technol.* **102** 1 2–9

[3] Rashid N, Ur Rehman M S, Sadiq M, Mahmood T and Han J-I 2014 Current status, issues and developments in microalgae derived biodiesel production *Renew. Sustain. Energy Rev.* **40** 760–78

[4] Singh A, Nigam P S and Murphy J D 2011 Mechanism and challenges in commercialisation of algal biofuels *Bioresour. Technol.* **102** 26–34

[5] Xiao R, Chen R, Zhang H-Y and Li H 2011 Microalgae *Scenedesmus quadricauda* grown in digested wastewater for simultaneous CO_2 fixation and nutrient removal *J. Biobased Mater. Bioenergy* **5** 2 234–40

[6] Rahman A and Miller C D 2017 Microalgae as a source of bioplastics. *Algal Green Chem* 121–38

[7] Aziz N, Fatma T, Varma A and Prasad R 2014 Biogenic synthesis of silver nanoparticles using *Scenedesmus abundans* and evaluation of their antibacterial activity *J. Nanoparticles* 1–6

[8] Mamo G and Mattiasson B 2019 Alkaliphiles: The emerging biological tools enhancing concrete durability *Advances in Biochemical Engineering/Biotechnology* (Berlin, Heidelberg: Springer) pp 1–50

[9] Joana Gil-Chávez G, Villa J A, Fernando Ayala-Zavala J, Basilio Heredia J, Sepulveda D, Yahia E M et al. 2013 Technologies for extraction and production of bioactive compounds to be used as nutraceuticals and food ingredients: an overview. *Compr. Rev. Food Sci. Food Saf.* **12** 5–23

[10] Molina Grima E, Belarbi E-H, Acién Fernández F, Robles Medina A and Chisti Y 2003 Recovery of microalgal biomass and metabolites: process options and economics. *Biotechnol. Adv.* **20** (7–8) 491–515

[11] Durakovic B 2017 Design of experiments application, concepts, examples: state of the art *Period. Eng. Nat. Sci.* **5** 3 421–39

[12] Gorman D S and Levine R P 1965 Cytochrome f and plastocyanin: their sequence in the photosynthetic electron transport chain of *Chlamydomonas reinhardi Proc. Natl. Acad. Sci. U S A* **54** 6 1665–1669

[13] Baranyi J and Roberts T A 1994 A dynamic approach to predicting bacterial growth in food *Int. J. Food Microbiol.* **23** (3–4) 277–94

[14] Zhang X, Amendola P, Hewson J C, Sommerfeld M and Hu Q 2012 Influence of growth phase on harvesting of *Chlorella zofingiensis* by dissolved air flotation *Bioresour. Technol.* **116** 477–84

[15] Qian Q, Zhang J, Cui M and Han B 2016 Synthesis of acetic acid via methanol hydrocarboxylation with CO_2 and H_2 *Nat. Commun.* **7** 11481

[16] Sakamoto T and Bryant D A 1997 Growth at low temperature causes nitrogen limitation in the cyanobacterium Synechococcus sp. PCC 7002 *Arch. Microbiol.* **169** 10–9

Acknowledgment

This study has been funded by Universiti Putra Malaysia (Putra Grant no.: GP-IPM/2017/9551600)

IOP Conf. Series: Materials Science and Engineering **716** (2020) 012004 doi:10.1088/1757-899X/716/1/012004

Anti-diabetic properties of *Synsepalum dulcificum* and its potential inclusion in functional yogurt

N F Fazilah[1], A B Ariff[1,2], M E Khayat[2,3], M Halim[1,2,]

[1]Department of Bioprocess Technology, Faculty of Biotechnology and Biomolecular Sciences, Universiti Putra Malaysia, 43400 UPM Serdang, Selangor, Malaysia
[2]Bioprocessing and Biomanufacturing Research Centre, Faculty of Biotechnology and Biomolecular Sciences, Universiti Putra Malaysia, 43400 UPM Serdang, Selangor, Malaysia
[3]Department of Biochemistry, Faculty of Biotechnology and Biomolecular Sciences, Universiti Putra Malaysia, 43400 UPM Serdang, Selangor, Malaysia

Email: murnihalim@upm.edu.my

Abstract. There has been an enormous interest in the development of alternative medicines for type-2 diabetes, specifically screening for phythochemicals with the ability to delay or prevent glucose absorption. The goals of the present study were to provide *in vitro* evidence for potential inhibition of α- amylase and α- glucosidase enzymes, followed by inclusion of the extracts of *Synsepalum dulcificum* in yogurt to enhance the therapeutic properties of the yogurt. The screening results of seed, leaf and pulp of *S. dulcificum* showed that pulp extracts contained significantly ($P<0.05$) higher anti-diabetic activities than the other plant parts. More interestingly, *S. dulcificum* pulp also has stronger anti-diabetic properties than the standard drug, acarbose and hence it was chosen to be incorporated into yogurt. *S. dulcificum* yogurt had higher ($P<0.05$) anti-diabetic activities than the plain yogurt throughout the storage period with the highest α-glucosidase and α-amylase inhibitory activities were shown on day 7 of storage. Therefore, *S. dulcificum* pulp can be developed as functional factor with anti-diabetic activities in food application.

1. Introduction

In this century, it is estimated at least 346 million of people worldwide suffer from type-2 diabetes mellitus due to pancreatic β-cell dysfunction and/or increased resistance to insulin with impaired glucose tolerance [1]. An effective strategy for type-2 diabetes management is the inhibition of α-amylase and α-glucosidase. α- amylase is an enzyme found in the pancreatic juice and saliva functions to break down large insoluble starch molecules into absorbable molecules [2]. In the meantime, α-glucosidase in the mucosal brush border of the small intestine catalyzes the end step of digestion of starch and disaccharides that are abundant in human diet [3]. Inhibitors of both α-amylase and α-glucosidase will act to delay the breaking down of carbohydrates in the small intestine and diminish the postprandial blood glucose excursion [4]. Phenolic compounds are known to interact with proteins and can inhibit enzymatic activity. To date, many plant and herbal extracts have been reported to efficiently inhibit the enzymatic activity of α-amylase and α-glucosidase [4, 5].

Synsepalum dulcificum also known as miracle fruit or miracle berry is an indigenous tropical plant growing in West Africa. The plant can be harvested at least twice a year making the yield stable [6]. Miracle fruit is about an inch in length with a bright red colour and known for its ability in converting sour tasting foods to sweet taste. Glycoprotein known as miraculin found in the pulp of this fruit is the compound responsible for this unique taste modifying function [6, 7]. The binding of miraculin to the receptor cells of the tongue suppresses the response of a sour taste in the central nervous system. The effect would last until the miraculin is diluted and eliminated by saliva. The taste modification function gives a great potential for this fruit to be exploited in the food industry especially as an alternative sweetener or additive to mask the undesirable sour taste in food products such as yogurt [8]. Moreover, miracle fruit's pulp and seed contained high nutrient contents which can be used for dietary supplement in food [9]. Stem of miracle fruit also contained antioxidant and antityrosinase effects which can be potential applications in food supplementation and medical cosmetology [10]. Hence, it is interesting for miracle fruit to be formulated into functional foods such as yogurt to improve the nutritional and therapeutic properties [11].

To date, very little is known about the health benefits of *S. dulcificum* plant extract. The ability of this plant to inhibit the enzymatic activity of α-amylase and α-glucosidase when incorporated into food product also has yet to be studied and explored. Keeping in view the potential of anti-diabetic properties of *S. dulcificum* plant extracts [12], the present study is endeavoured to develop functional yogurts fortified with *S. dulcificum* extract. In this regard, the inclusion of the plant extracts is expected to enhance the nutritional values of the yogurt and as such functional food that contained components which are able to inhibit α-amylase and α-glucosidase could be developed.

2. Materials and methods

2.1. Plant materials

The pulp, seed and leaf of S. dulcificum were obtained from local farm located in Selangor. They were thoroughly washed with tap water to remove unwanted materials. Then, the seeds were separated from the pulp of S. dulcificum using a knife, and the pulp and seed were freeze dried for 48 hours by Martin Christ Epsilon 1-80 freeze dryer. The leaf samples were dried in an oven at 50°C for 24 hours. All the samples were then grinded using a household blender into fine powders and kept in labelled zip lock plastics in a chiller (4°C) until usage.

2.2. Starter culture

Yogurt starter culture was purchased from Chris Hansen (Denmark) in the form of freeze-dried containing a mixture of bacteria strains (YC-380): Streptococcus thermophilus and Lactobacillus delbrueckii subs. Bulgaricus in the ratio of 1:1. Briefly, the freeze-dried yogurt starter was inoculated in fresh pasteurized cow milk in a ratio of 1:100 (wv^{-1}). The mixture was then incubated at 42°C overnight without shaking. The yogurt formed was kept at 4°C and used as a starter culture within 1 week.

2.3. Water extraction of S. dulcificum plant

Water extraction of plant samples was done following a method by Amirdivani [5] with some modifications. The fine powders of pulp, seed and leaf were added in distilled water in a ratio of 1:10 (wv^{-1}). Then, the solutions were incubated in a water bath (50°C) for 16 hours, followed by filtration with Whatman's filter paper No. 5. The filtrate was harvested and centrifuged (5810R, Eppendorf, Germany) at 6000 g for 10 minutes. The pellet was discarded while the supernatant was stored in 4°C and used within 3 days.

2.4. Preparation of yogurt

Plain and *S. dulcificum* yogurts were prepared on the same day. *S. dulcificum* pulp extract (10 mL) was added into pre-warmed pasteurized cow milk (80 mL) followed by the addition of starter culture (10 g). Meanwhile, plain yogurt was prepared essentially in the same manner as the *S. dulcificum* yogurts with the exception that distilled water (10 mL) was used instead of the pulp extract. Yogurts were fermented (42°C) until its pH reduced to 4.5 followed by refrigeration (4°C) up to 21 days.

2.5. Preparation of yogurt extracts

Yogurt extracts were prepared following a method reported by Amirdivani [5] with some modifications. Aliquots of each 10 g yogurt sample was homogenized (by vortex) with 2.5 mL sterile distilled water. The pH of the yogurt was acidified to 4.0 with 0.1 M hydrochloric acid (HCl). The yogurts were then heated in a water bath (45°C) for 10 minutes before centrifuged (MiniSpin, Eppendorf, Germany) at 7500 g, 4°C for 10 minutes. The supernatants were harvested and pH was adjusted to neutral (pH 7) with an addition of 0.1 M sodium hydroxide (NaOH). The yogurt water extracts were re-centrifuged under the same condition and the supernatants were harvested and kept in -20°C prior to further analysis.

2.6. α-amylase inhibitory activity

Following a method described by Akoro *et al.*, [13] with some modifications, 250 μL enzyme solution (1 U/mL of pancreatic alpha-amylase enzyme (E.C:3.2.1.1; Sigma, USA) dissolved in 0.02 M sodium phosphate buffer with 0.006 M sodium chloride) was added to 250 μL sample (plant extract/ control) and incubated for 10 minutes at 37°C. Then, 250 μL of 1% soluble starch (potato starch) was added and incubated again for 10 minutes at 37°C. The reaction was terminated by adding 500 μL of DNSA (3,5-dinitrosalicylic acid) reagent and then boiled in the water bath at boiling point for 5 minutes. The solution was cooled and diluted with 5 mL of water and absorbance reading was taken at 540 nm. To eliminate the absorbance produced by plant extract, an appropriate extract control without the present of enzyme was also included. The results were expressed as percentage of inhibition of extracts against concentration, calculated according to the Equation 1:

$$\%\ inhibition = \frac{(Absorbance\ of\ control - Absorbance\ of\ extract)}{Absorbance\ of\ control}\ x\ 100 \quad (1)$$

2.7. α-glucosidase inhibition assay

The effect of *S. dulcificum* extracts on α-glucosidase activity was determined according to the method described by Kazeem *et al.*, [4] with some modifications, using α-glucosidase from *Saccharomyces cerevisiae*. The substrate solution p-nitropheynyl glucopyranoside (pNPG) (3.0 mM) was prepared in 20 mM phosphate buffer, pH 6.9. 100 μL of α-glucosidase (1.0 U/mL) was pre-incubated with 50 μL of the extracts for 10 minutes. Then 50 μL of 3.0 mM pNPG dissolved in 20 mM phosphate buffer (pH 6.9) was added to start the reaction. The reaction mixture was incubated at 37°C for 20 minutes and stopped by adding 2 mL of 0.1 M Na_2CO_3. The α-glucosidase activity was determined by measuring the yellow coloured para-nitrophenol released from pNPG at 405 nm. The results (% inhibition) were expressed as percentage of the blank (control) as in Equation 1.

2.8. Statistical analysis

All experiments were performed on at least three different occasions (n=3). All data is expressed as mean value ± standard deviation. Data analyses and the graphs were analyses and plotted on Graph Pad prism 6. The statistical significance was determined at p value < 0.05. Statistical analysis was done using the Analysis of variance (ANOVA); $P<0.05$ was considered significant.

3. Results and discussion

3.1. Screening of anti-diabetic properties in S. dulcificum plant parts

3.1.1. α-amylase inhibitory activity

α-amylase catalyzes the hydrolysis of 1,4-glycosidic linkages of starch, glycogen, and various oligosaccharides into simpler sugars which can be readily available for the intestinal absorption [14]. Therefore, inhibition of this enzyme in the digestive tract of human is considered to be effective in controlling diabetes by decreasing the absorption of glucose from starch [15]. In this study, porcine pancreatic α-amylase (PPA) was used since PPA is closely related to human α-amylase [16]. *In vitro* α-amylase inhibitory studies demonstrated that all the *S. dulcificum* extracts possessed α-amylase inhibitory activity (Table 1).

The evaluation of α-amylase inhibitory activity of acarbose (an anti-diabetic drug used to treat diabetes mellitus type 2) was also done to assess the validity of the assay used. The results revealed that a dose-dependent response ($P<0.05$) since increasing concentration of plant extracts increased the inhibitory activity (Table 1). At the lowest concentration of extracts (1.25 mg/mL), the pulp extract exhibited the highest inhibitory activity of 20.73% which was higher than that of acarbose (13.4%). However, at the highest concentration (20 mg/mL), the highest inhibition was observed in acarbose (78.35%) with leaf extract having the lowest α-amylase inhibitory activity of 31.96%. Nevertheless, the overall result showed that the *S. dulcificum* pulp extract is the most effective for the inhibitory of α-amylase activity since the pulp was having the lowest IC_{50} value of 9.66 mg/mL when compared to acarbose and the other two extracts. This result indicated that the *S. dulcificum* pulp is a strong α-amylase inhibitor that could help to suppress the hydrolysis of starch leading to reduction in post-prandial glucose level and hence reducing the complication of diabetes mellitus. The presence of α-amylase inhibitor in *S. dulcificum* pulp may be due to the presence of potential α-amylase inhibitors such alkaloids, flavonoids, terpenoids or glycosides [17]. To our knowledge till date, the present study is the first to report on α-amylase inhibitory activity in *S. dulcificum*.

Table 1. α-amylase inhibitory activity of *S. dulcificum* plant parts and synthetic drug, acarbose

Concentration (mg/mL)	**% Inhibition by seed**	**% Inhibition by leaf**	**% Inhibition by acarbose**	**% Inhibition by pulp**
1.25	$18.56\pm(0.01)^{a}$	$4.12\pm(0.00)^{a}$	$13.4\pm(0.00)^{a}$	$20.73\pm(0.01)^{a}$
2.50	$20.62\pm(0.03)^{b}$	$18.56\pm(0.01)^{b}$	$28.87\pm(0.02)^{b}$	$34.14\pm(0.02)^{b}$
5.00	$20.62\pm(0.01)^{c}$	$19.59\pm(0.01)^{c}$	$41.24\pm(0.01)^{c}$	$39.02\pm(0.02)^{c}$
10.00	$22.68\pm(0.02)^{d}$	$21.65\pm(0.01)^{d}$	$47.42\pm(0.01)^{d}$	$59.76\pm(0.01)^{d}$
20.00	$35.05\pm(0.00)^{e}$	$31.96\pm(0.06)^{e}$	$78.35\pm(0.02)^{e}$	$71.95\pm(0.02)^{e}$
IC_{50} value (mg/mL)	39.37	35.09	10.4	9.66

Data was expressed as mean ± standard deviation of triplicate determinations.
IC_{50}: Concentration of inhibitor to inhibit 50% of the enzyme activity under the assayed conditions.
[a,b,c,d,e] Means values in the same column expressed with different superscript letters are significantly different at ($P< 0.05$)

3.1.2. α-glucosidase inhibitory activity

Glucosidase is the key enzyme involved in hydrolysis of carbohydrate and inhibitors of this enzyme may be exploited as therapeutic approaches for controlling postprandial hyperglycemia [18]. Phytochemicals from plants such as phenolic compounds, saponins, flavonoids, glycosides and alkaloids have been reported to play an important role in modulating glucosidase activity and can therefore contribute to the management of postprandial hyperglycemia [19].

From the result obtained, a similar trend as α-amylase inhibitory activity result was observed in which increasing the concentration of plant extracts resulted in increasing of α-glucosidase inhibitory activity (Table 2). At the highest concentration of 0.5 mg/mL, pulp extract showed a maximum and effective inhibition of nearly 98.94 % whereas acarbose (control), showed a lower inhibition of 40.10%, under similar assay condition. The IC_{50} of *S. dulcificum* pulp, seed and leaf were 0.08, 0.23, 0.30 mg/mL, respectively, which were evidently less than that for acarbose (0.70 mg/mL). The lower value of IC_{50} in *S. dulcificum* extracts showed that all the extracts are effective and strong inhibitors of α-glucosidase when compared to the synthetic drug, acarbose.

When comparing the results of α-amylase IC_{50} values to α-glucosidase, all the *S. dulcificum* extracts showed higher α-glucosidase inhibitory activity than α-amylase. According to Mogela *et al.*, [20] natural inhibitors from plants are mostly reported to have a lower inhibitory effect against α-amylase but stronger inhibitory activity against α-glucosidase. Anuya and Abhay, [21] also found similar results whereby the aqueous extract of *Ocimum sanctum Linn* showed lower α-amylase inhibitory activity and stronger α-glucosidase inhibitory activity than the ethanol extract.

Overall, due to the strong inhibitory properties, *S. dulcificum* pulp has great potential for use in functional food application and thus pulp extract was selected for the subsequent experiments of the yogurt preparation. Additionally, the strong α-glucosidase inhibition and mild inhibition of α-amylase makes the extract a good source for management of type-2 diabetes with potentially minimum side effects than that currently observed with several drugs being used for management of non-insulin dependent diabetes mellitus. For example, the use of acarbose is reported to be associated with gastrointestinal side effects caused by the excessive inhibition of pancreatic α-amylase, resulting in the abnormal bacterial fermentation of undigested carbohydrates in the large intestine [22].

Table 2. α-glucosidase inhibitory activity of *S. dulcificum* plant parts and synthetic drug, acarbose

Concentration (mg/mL)	% Inhibition by acarbose	% Inhibition by leaf	% Inhibition by seed	% Inhibition by pulp
0.03	10.35±(0.05)^{a}	11.20±(0.02)^{a}	17.90±(0.01)^{a}	20.92±(0.01)^{a}
0.06	11.04±(0.02)^{b}	23.10±(0.01)^{b}	26.57±(0.05)^{b}	48.49±(0.03)^{b}
0.13	15.37±(0.01)^{c}	31.60±(0.05)^{c}	35.34±(0.01)^{c}	68.73±(0.02)^{c}
0.25	16.80±(0.01)^{d}	45.90±(0.03)^{d}	54.89±(0.02)^{d}	92.34±(0.01)^{d}
0.50	40.10±(0.02)^{e}	72.80±(0.07)^{e}	86.63±(0.04)^{e}	98.94±(0.06)^{e}
IC_{50} value (mg/mL)	0.70	0.30	0.23	0.08

Data was expressed as mean ± standard deviation of triplicate determinations.
IC_{50}: Concentration of inhibitor to inhibit 50% of the enzyme activity under the assayed conditions.
^{a,b,c,d,e} Means values in the same column expressed with different superscript letters are significantly different at ($P < 0.05$).

3.2 Formulation of functional yogurt

3.2.1. α-amylase inhibitory activity

In this study, the α-amylase inhibitory activities increased significantly ($P < 0.05$) for both types of yogurts after the fermentation step, up to day 7 (Figure 1). At day 7, which is the highest α-amylase inhibitory activities, high inhibition of 69.23% was observed in *S. dulcificum* yogurt when compared to plain yogurt (18.46%). According to Ni *et al.*, (2018), the increase of α-amylase inhibitory activities in yogurts supplemented with salal berry (*Gaultheria shallon*) and blackcurrant (*Ribes nigrum*) pomace during cold storage were attributed to the viability of lactic acid bacteria. The proteolytic activity of lactic acid bacteria [23] contributed to the release of peptides with moderate α-amylase inhibitory activity.

Refrigerated storage also has profound effect on yogurts inhibition of α-amylase activity. This was demonstrated by a decrease in α-amylase inhibition ($P < 0.05$) about 26% during storage from day 7 to day 21. The present study showed that the inclusion of *S. dulcificum* pulp extract in yogurt could increase the therapeutically values of the yogurt via higher α-amylase inhibitory activity during storage compared to plain yogurt.

3.2.2. α-glucosidase inhibitory activity

The α-glucosidase inhibitory activity was significantly ($P < 0.05$) affected by both types of yogurt and storage duration as shown in Figure 2. Yogurt that supplemented with *S. dulcificum* pulp had higher activity than the plain yogurt with the highest α-glucosidase inhibitory activity was observed on day 7. High α-glucosidase inhibitory activity of yogurt fortified with *S. dulcificum* pulp indicated that the addition of pulp extract enhanced the inhibitory activity. This could be mainly due to the presence of phytochemical compounds in the pulp extract such as anthocyanin, flavonoids and phenolic compounds [6]. Anthocyanin is sub-type of organic compound of flavonoid family and a member of large group of compounds [24]. *S. dulcificum* was previously reported to have 11.04 mg/100 g FW of total anthocyanin content, that possibly responsible for the red color of the fruit [6].

However, when compared to the screening result, there was high reduction of inhibitory activity of this enzyme when fortified into yogurt. This could be due to the low content of flavonol compounds in the yogurt. Flavonol compounds, such as quercetins and its derivatives have particularly high affinity towards the α-glucosidase enzyme *in vitro* [25]. Additionally, the decreased of the activity as observed after day 7 might be due to the anthocyanin stability in the *S. dulcificum* pulp extract. Arueya and Akomolafe [26], stated that anthocyanin are highly unstable molecules in food matrix and their stability is strongly affected by pH, solvents, temperature, oxygen, light, enzymes, anthocyanin concentration, structure and other accompanying substances. Their studies also found similar results where there was a decrease in α-glucosidase inhibitory activity of yogurt fortified with *Allium sativum* from 15.2% to 12.8% during 21 days of cold storage. Likewise, the α-glucosidase inhibitory activity of probiotic yogurt supplemented with Roselle extract was decreased during 15 days of cold storage [27].

On other hand, for plain yogurt, the inhibitory activity observed was most probably due to the presence of bioactive peptides produced through proteolysis by LAB [28]. During fermentation, protein content in the milk is degraded by LAB into peptides. Several bioactive peptides have been found to be able to inhibit the activity of α-glucosidase enzyme [29]. Moreover, probiotic strain also has been reported to have an anti-diabetic effect. For instant, *Lactobacilli* probiotic has the potency to be anti-diabetic probiotic due to its ability to act as α-glucosidase inhibitor [30].

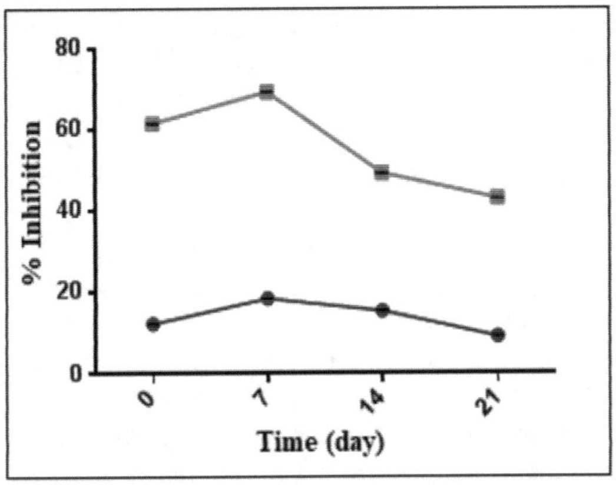

Figure 1. α-amylase inhibitory activity of yogurt fortified with *S. dulcificum* pulp extract (■) and plain yogurt (●) during cold storage at 4°C. The error bars represent the standard deviations about the mean (*n*=3).

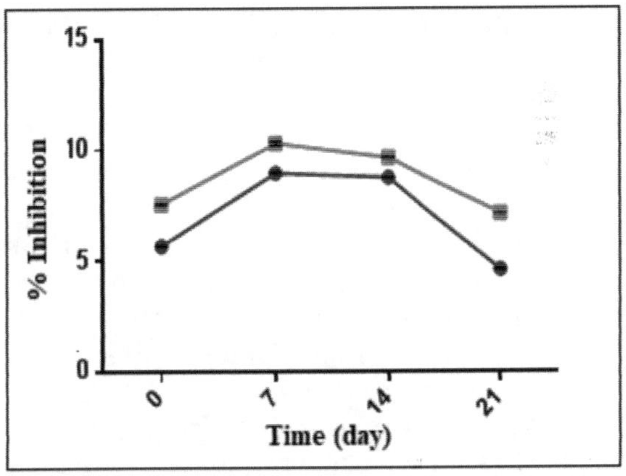

Figure 2. α-glucosidase inhibitory activity of yogurt fortified with *S. dulcificum* pulp extract (■) and plain yogurt (●) during cold storage at 4°C. The error bars represent the standard deviations about the mean (*n*=3).

4. Conclusion

Overall, the data reported in this study suggesting that *S. dulcificum* pulp extract contained high α-amylase and α-glucosidase inhibitory activities. *S. dulcificum* pulp showed stronger α-amylase (IC_{50} 9.66) and α-glucosidase (IC_{50} 0.08) inhibitory activities compared to standard drug, acarbose, seed and leaf. In the view of these traits, the consumption of *S. dulcificum* pulp could be of great value in prevention of oxidative stress related diseases (such as antherosclerosis, coronary heart diseases and cancer) and may also be useful in the management of diabetes. The inclusion of *S. dulcificum* pulp extract into yogurt has the potential to be a new functional food-based product with high nutritional appeal and beneficial properties. However, from this preliminary study it was found that the α-glucosidase inhibitory activity showed a gradual decrement when the extract was formulated into yogurt, thus it is interesting to identify the compounds that are responsible on affecting the interactions between food matrix and α-glucosidase enzyme.

IOP Conf. Series: Materials Science and Engineering **716** (2020) 012004 doi:10.1088/1757-899X/716/1/012004

References

[1] Danaei G, Finucane M, Lu Y, Singh G, Cowan M, Paciorek C, Stevens G and Global Burden of Metabolic Risk Factors of Chronic Diseases Collaborating Group (Blood Glucose) 2011 National, regional, and global trends in fasting plasma glucose and diabetes prevalence since 1980: systematic analysis of health examination surveys and epidemiological studies with 370 country-years and 2.7 million participants *Lancet* **378(9785)** 31-40

[2] Afifi A F, Kamel E M, Khalil A A, Foaad M A, Fawziand E M and Houseny M 2008 Purification and characterization of a-amylase from *Penicillium olsonii* under the effect of some antioxidant vitamins *Global Journal of Biotechnology and Biochemistry* **3(1)**14–12.

[3] Manohar V, Talpur N A, Echard B W, Lieberman S and Preuss H G 2002 Effects of a water-soluble extract of maitake mushroom on circulating glucose/insulin concentrations in KK mice *Diabetes, Obesity and Metabolism* **4(1)** 43–48

[4] Kazeem M I, Adamson J O and Ogunwande I A 2013 Modes of inhibition of α-amylase and α-glucosidase by aqueous extract of *Morinda lucida* benth leaf *Biomed Research International* 1-6

[5] Amirdivani S 2015 Inclusion of *Allium sativum* in yogurt and its effects on inhibition of diabetes and hypertension-associated by enzymes in vitro *Applied Food Biotechnology* **2(3)** 29-37

[6] He Z, Tan J S, Abbasiliasi S, Lai O M and Ariff A B 2016 Phytochemicals, nutritionals and antioxidant properties of miracle fruit *Synsepalum dulcificum Industrial Crops and Products* **86** 87-94

[7] He Z, Tan J S, Abbasiliasi S, Lai O M, Tam Y J, Halim M and Ariff A B 2015 Primary recovery of miraculin from miracle fruit, *Synsepalum dulcificum* by AOT reverse micellar system *LWT-Food Science and Technology* **64** 1243-1250

[8] Wilken M K ,and Satiroff B A 2012 Pilot study of miracle fruit to improve food palatability for patients receiving chemotherapy *Clinical Journal of Oncology Nurs*ing **16** 163-166

[9] Agblekpe A K, Osseyi E and Dossou J 2016 Potential nutritional values of skin, pulp and seed of miracle fruit (*Synsepalum dulcifium*) *American Journal of Innovative Research and Applied Sciences* **4(1)** 1–7

[10] Wang H M, Chou Y T and Hong Z L 2011 Bioconstituents from stems of *Synsepalum dulcificum* Daniell (Sapotaceae) inhibit human melanoma proliferation, reduce mushroom tyrosinase activity and have antioxidant properties *Journal of the Taiwan Institute of Chemical Engineers* **42(2)** 204–211

[11] Fazilah N F, Ariff A B, Khayat M E, Rios-Solis L and Halim M 2018 Influence of probiotics, prebiotics, synbiotics and bioactive phytochemicals on the formulation of functional yogurt *Journal of Functional Foods* **48** 387-399

[12] Chen C C, Liu I M and Cheng J T 2006 Improvement of insulin resistance by miracle fruit (*Synsepalum dulcificum*) in fructose rich chow-fed rats *Phytotherapy Research* **20(11)** 987–992

[13] Akoro S M, Ogundare C O and Oladipupo O R 2017 Phytochemical study and alpha-amylase inhibitory properties of leaf extracts of *Croton zambesicus* (Müll. Arg.) *Biotechnology Journal International* **18(1)** 1-8

[14] Sundarram A and Murthy T P K 2014 α – Amylase production and applications: A review. *Journal of Applied & Environmental Microbiology* **2(4)** 166-175

[15] Saeidnia S, Ara L, Hajimehdipoor H, Roger W, Read R W, Arshadi S and Nikan M 2016 Chemical constituents of *Swertia longifolia* Boiss with α -amylase inhibitory activity *Research in Pharmaceutical Sciences* **11(1)** 23-32

[16] Sudha P, Zinjarde S S, Bhargava S and Kumar A R 2011 Potent α-amylase inhibitory activity of Indian Ayurvedic medicinal plants *BMC Complementary and Alternative Medicine* **11** 5

[17] Ponnusamy S, Ravindran R, Zinjarde S, Bhargava S and Kumar A R 2011 Evaluation of traditional Indian antidiabetic medicinal plants for human pancreatic amylase inhibitory

eggect in vitro *Evidence-Based Complentary and Alternative Medicine* doi:10.1155/2011/515647

[18] Bhat M, Zinjarde S S, Bhargava S Y, Kumar A R and Joshi B N 2011 Antidiabetic Indian plants: A good source of potent amylase inhibitors *Evidence-Based Complementary Alternative Medicine* doi: 10.1093/ecam/nen040

[19] Ani V and Akhilender N K 2008 Antihyperglycemic activity of polyphenolic components of black/ bitter cumin *Centratherum anthelminticum* (L.) Kuntze seeds. *Europian Food Resource Technology* **226** 897-903

[20] Mogale M A, Lebelo S L, Thovhogi N, de Freitas A N and Shai L J 2011 α-amylase and α-glucosidase inhibitory effects of *Sclerocarya birrea* [(A. Rich.) Hochst.] subspecies caffra (Sond) Kokwaro (Anacardiaceae) stem-bark extracts. *African Journal of Biotechnology* **10** 15033-15039

[21] Anuya A A R and Abhay S C 2014 Evaluation of alpha-amylase and alpha-glucosidase inhibitory activities of *Ocimum sanctum* Linn *International Journal Pharmacology Science* **25**(**1**) 130-133

[22] Rosak C and Mertes G 2012 Critical evaluation of the role of acarbose in the treatment of diabetes: patient considerations *Diabetes, Metabolic Syndrome and Obesity: Targets and Therapy* **5** 357-367

[23] Othman M, Ariff A B, Wasoh H, Kapri M R, and Halim M 2017 Strategies for improving production performance of probiotic *Pediococcus acidilactici* viable cell by overcoming lactic acid inhibition AMB Express **7** 215 doi:10.1186/s13568-017-0519-6

[24] Skrovankova S, Sumczynski D, Mlcek J, Jurikova T and Sochor J 2015 Bioactive compounds and antioxidant activity in different types of berries *International Journal of Molecular Sciences* **16**(**10**) 24673-24706

[25] Kongstad K T, Özdemir C, Barzak A, Wubshet S G and Staerk D 2015 Combined used of high-resolution α-glucosidase inhibition profiling and high-performance liquid chromatography-high-resolution mass spectrometry-solid-phase extraction-nuclear magnetic resonance spectroscopy for investigation of antidiabetic principles in crude plant extracts *Journal of Agricultural and Food Chemistry* **63**(**8**) 2257-2263

[26] Arueya G L and Akomolafe B O 2014 Stability studies of microencapsulated anthocyanins of roselle (*Hibiscus sabdar¬iffa L.*) in native starch and its potential application in jam production *IOSR Journal of Environmental Science, Toxicology and Food Technology* **8** 112-122

[27] Wihansaha R R S, Arief I I and Batubarac I 2018 Anti-diabetic potency and characteristics of probiotic goat-milk yogurt supplemented with roselle extract during cold storage *Animal Science Journal* **41**(**3**)191-199

[28] Ming L C, Halim M, Abd Rahim R, Wan H Y and Ariff A B 2016 Strategies in fed-batch cultivation on the production performance of *Lactobacillus salivarius* I 24 viable cells *Food Science and Biotechnology* **25**(**5**) 1393-1398

[29] Lacroix I M and Li-Chan E C Y 2013 Inhibition of dipeptidyl peptidase (DPP)-IV and α-glucosidase activities by pep¬sintreated whey proteins *Agriculture Food Chem*istry **61** 7500–7506

[30] Zeng Z, Luo J, Zuo F, Zhang Y, Ma H and Chen S 2016 Screening for potential novel probiotic *Lactobacillus* strains based on high dipeptidyl peptidase iv and a-glucosidase inhibitory activity *Journal of Functional Foods* **20** 486-495

Acknowledgement
This research was funded by Universiti Putra Malaysia, grant number 9559400 (Geran Putra IPS).

Fermentation of *Dioscorea hispida* oligosaccharides as prebiotic potential by lactic acid bacteria and bifidobacterium

H Napisah[1] and A Rosma[2,3]
[1] Faculty of Health Sciences, Kampus Gong Badak, Universiti Sultan Zainal Abidin, 21300 Kuala Terengganu, Terengganu Darul Iman, Malaysia
[2] Bioprocess Technology Division, School of Industrial Technology, Universiti Sains Malaysia, 11800 USM, Penang, Malaysia

[3] Email: rosmah@usm.my

Abstract. *Dioscorea hispida* oligosaccharides (DHOS) were intended to be used as a synbiotic product. Fermentability of DHOS by lactic acid bacteria and Bifidobacterium was studied. *In vitro* DHOS fermentations by *L. acidophilus* FTDC 2131, *L. casei* ATCC 393, *L. casei* FTCC 0442, *L. plantarum* FTCC 0350 and *B. bifidum* 12 were at 37°C for 24 h. DHOS was the most preferred substrate by *L. acidophilus* FTDC 2131, *L. plantarum* FTCC 0350 and *B. bifidum* 12, whilst, fructooligosaccharides (FOS) and inulin was the best substrate for the growth of *L. casei* ATCC 393 and *L. casei* FTCC 0442, respectively. The pH reduction in DHOS fermentation medium was comparable to the commercial FOS suggesting that DHOS was able to enhance the selected bacterial growth and produced organic acids. Fermentation of DHOS by *B. bifidum* 12 had exhibited the highest prebiotic activity score (+0.713) and produced 30.42 mM lactic, 22.63 mM acetic and 0.71 mM propionic acids. The incorporation of DHOS in the medium culture of the selected bacteria was able to enhance the bacterial growth, reduce the pH medium, produce substantial amount of lactic, acetic and propionic acids, and exhibit positive prebiotic activity score.

1. Introduction

Dioscorea hispida belongs to Dioscoreaceae family which has about 1137 species with only 600 species are edible [1]. *D. hispida* is a monocotyledous perennial climbing plant that produces tubers. However, the tubers contain water soluble toxic alkaloid dioscorine. Consequently, *D. hispida* tubers have low economic value, although a plant of *D. hispida* can generally produce 15 kg of mature tubers. However, sometimes the plant can produce up to 25-30 kg of tubers depending on the climate. Each 100 g of tubers could produce up to 20 g of starch [2]. *D. hispida* tuber is synonymous with poor people daily food as well as an important food during famine [3]. *D. hispida* is widely found as a wild plant in tropical and subtropical regions of the world. Besides the tubers, *D. hispida* plant has significant benefit in ethnopharmaceutical practice as the plant extract has been traditionally used in Bangladesh to treat arthritis, ulcer and skin infections [4]. It was also revealed that the bioactive extracts of *D. hispida* plant had antioxidant, anti-microbial, thrombolytic and membrane stabilization activities [4].

Tubers and roots which generally do not contain gluten that causes incidence of celiac disease or allergic reactions are important sources of carbohydrates in tropical and subtropical countries as staple foods [5]. Therefore, attempts are being made by the Malaysian government to breed and cultivate the *D. hispida* plant for its tubers in large scale to fulfil the government food security agenda [6].

Properties of *D. hispida* starch are unique which can be used as a new source of starch for various applications [7]. Conventionally, *D. hispida* tubers are submerged in free flowing water such as river stream or submerged in large volume of contained water for 3-5 days with daily change of water to remove the dioscorine. Usually, the tubers are consumed together with grated coconut or dipping sauce after boiling, steaming or frying the tubers [2, 3].

Starch of *D. hispida* tuber could be value added by hydrolysing the starch into oligosaccharides. Oligosaccharides are a group of carbohydrate polymers with degree of polymerisation (DP) of 2 to 12. The oligosaccharides are regarded as prebiotic if they are resistant to digestion and absorption in the human small intestine but fermented into short chain fatty acids by beneficial intestinal microflora in the large intestine [8]. The US National Centre for Complementary and Integrative Health refers prebiotic as dietary substance that favours the growth of beneficial microbes over the harmful ones [9]. The Centre also refers synbiotic as a product that contains both prebiotic and probiotic. The US Food and Drug Agency only recognizes inulin, fructooligosaccharides, galactosyloligosaccharides and xylooligosacchradides from various sources as prebiotics although lactulose, mannitol, maltodextrin, raffinose and sorbitol are also commercially used as prebiotics [10]. In this research, starch obtained from D. hispida tuber was further processed via enzymatic hydrolysis to oligosaccharides. Experiments were conducted to evaluate the prebiotic activity of *D. hispida* oligosaccharides by selected lactic acid bacteria (*L. acidophilus* FTDC2131, *L. casei* ATCC39, *L. casei* FTCC0442, *L. plantarum* FTCC0350) and *B. bifidum* 12 against gut pathogens (*E. coli* FTDC201, *C. coli* c519, *C. perfringens* c438).

2. Materials and methods

2.1. Materials

D. hispida tubers were obtained from Marang, Terengganu, Malaysia. Commercial prebiotics FOS (Raftilose, Orafti® P95, BENEO-Orafti) and inulin (Raftiline, Orafti® GR, BENEO-Orafti) were used as positive controls were provided by DPO (M) Pvt. Ltd. (Kuala Lumpur, Malaysia).

2.2. Bacterial strains

Pure cultures of lactic acid bacteria were obtained from Bioprocess Technology Division (School of Industrial Technology, Universiti Sains Malaysia, Penang, Malaysia) and Food Technology Research Centre (Malaysian Agricultural Research and Development Institute, Selangor, Malaysia). *Bifidobacterium bifidum* 12 was provided by Chr-Hansen A/S (Malaysia Regional Office, Selangor, Malaysia) and pathogens (*Escherichia coli* FTDC 201, *Campylobacter coli* c519 and *Clostridium perfringens* c438) were obtained from the Institute for Medical Research, Kuala Lumpur, Malaysia. Lactic acid bacteria and bifidobacterium cultures were maintained on de Man, Rogosa, Sharpe agar (Hi-Media, India). *E. coli* and *C. coli* were maintained on the tryptic soy agar (Hi-Media, India), while *C. perfringens* was on Wilkins Chalgren anaerobic agar (Hi-Media, India).

2.3. Preparation of D. hispida tuber flour and starch samples

D. hispida tubers (total weight of 15 kg) were peeled, washed and cut into slices before submerged in excessive water for 4 days with daily constant change of water. Then, the slices were dried in an oven (Memmert UL40) at 50°C until a constant weight was obtained. The dried tuber slices were grinded and sieved to obtain flour with particle size of less than 180 μm. Starch was extracted from *D. hispida* tuber flour according to Hoover and Senanayake [11]. The starch was ground to particle size of 180 μm and stored in an airtight container until used.

2.4. Dioscorea hispida oligosaccharides (DHOS)

DHOS was prepared by the enzymatic hydrolysis of *D. hispida* starch. Initially, the starch slurry (10%, w/v) in 0.05 M citrate phosphate buffer, pH 5 was gelatinized at 100°C for 30 min [12]. Then, 5 mL of

IOP Conf. Series: Materials Science and Engineering **716** (2020) 012005 doi:10.1088/1757-899X/716/1/012005

50 ppm $CaCl_2$ was added, followed by 0.15 mL enzyme solution which consist of 25 GNU α-amylase and 125 NPUN pullulanase (Novozyme, Bagsvaerd, Denmark) and the sample mixture was incubated in a shaking water bath of 150 rpm for 12 h. The enzymatic hydrolysis was deactivated by boiling the samples for 10 min. The samples were centrifuged at 2500× g for 15 min and supernatant was transferred into clean tubes and analysed for reducing sugars (dinitrosalycylic acid method) and oligosaccharides concentration (HPLC). Maltooligosaccharides standard (Supelco 4-7265, USA) was used to estimate the concentration of *D. hispida* oligosaccharides.

2.5. In vitro fermentation and determination of cell viability
For lactic acid bacteria and bifidobacterium, fermentation medium was de Man, Rogosa, Sharpe broth except that the dextrose was replaced with 1% (w/v) of *D. hispida* oligosaccharides, FOS or inulin. Tryptic soy broth was for both *E. coli* and *C. coli* whereas Wilkin Chalgren anaerobic broth was for *C. perfringens*. Fermentation process was carried out at 37°C (in aerobic or anaerobic condition according to the required oxygen of the bacteria) for 24-72 h after inoculating with 1 mL bacterial inoculum (10^6 colony-forming unit, CFU) in 20 mL medium. All preparation, incubation of pre-culture and fermentation, sampling, serial dilution and agar plating for *B. bifidum* and *C. perfringens* (obligate anaerobic cultures) were carried out in an anaerobic chamber supplied with an atmospheric system of 85% N_2, 10% CO_2 and 5% H_2. The viable counts were determined by standard spread plate method incubated for 24-48 h and referred as $\log_{10}$ CFU/mL.

2.6. Prebiotics activity score
The prebiotic activity score (PAS) was calculated according to Huebner et al [13]:

$$\text{PAS} = \left(\frac{\text{Cell density changes on prebiotic}}{\text{Cell density changes on glucose}}\right) - \left(\frac{\text{Cell density changes of pathogen on prebiotic}}{\text{Cell density changes of pathogen on glucose}}\right) \quad (1)$$

The changes in cell density were the differences between the viable count at 0 h and 24 h (in $\log_{10}$ CFU/mL).

2.7. Determination of lactic and short chain fatty acids
Cell free supernatants (20 μL) were filtered and injected into a high performance liquid chromatographer (Shimadzu, Kyoto, Japan), which was attached with IC pak-ion exclusion column, 300 mm × 7.8 mm (Waters, USA) and UV-Vis detector (Shimadzu Model SPD-20A, Kyoto, Japan) at 210 nm. The elution took place at 50°C with 5 mM degassed H_2SO_4 at a flow rate of 0.6 mL/min. Concentrations of organic acids in the samples were referred to standard calibration curves of lactic, acetic and propionic acids (HPLC grade, Fluka, Steinheim, Germany).

2.8. Statistical analysis
The triplicate data was subjected to a one-way analysis of variance (ANOVA), and the significance of the differences between means was determined by the Duncan test, where $p < 0.05$ was considered statistically significant. The Statistical Package for Social Science, version 14.0 (SPSS Inc., U.S.A) was used for the analysis.

3. Results and discussion

3.1. Dioscorea hispida oligosaccharides
Figure 1 shows the HPLC chromatogram of *D. hispida* oligosaccharides. It contains a mixture of carbohydrate polymers of maltose (DP 2) to maltooligosaccharides with DP 10. The mixture also contained glucose but the amount was insignificant compared to the amount of oligosaccharides.

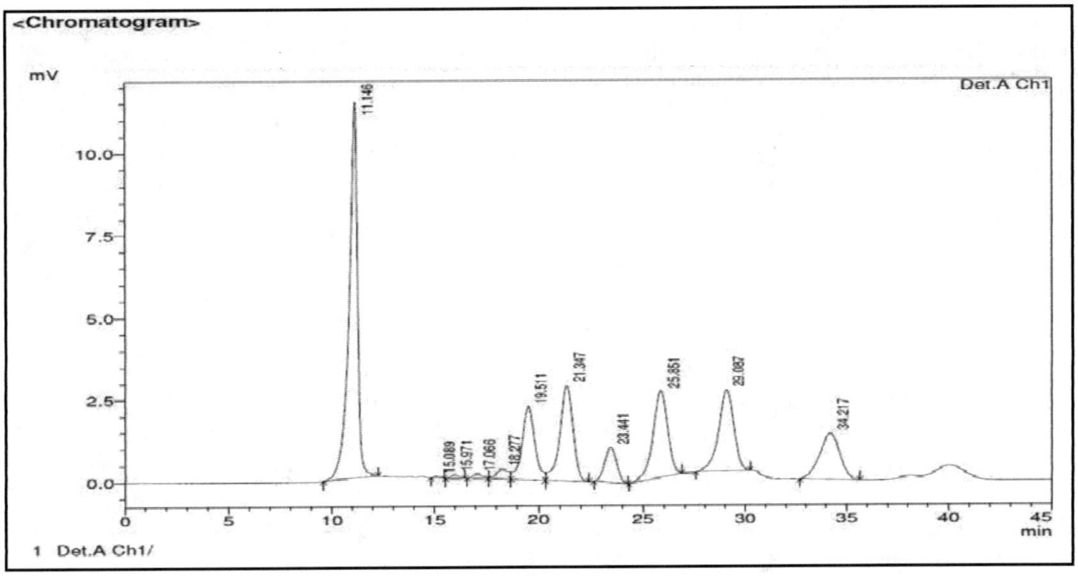

Figure 1. HPLC chromatogram of *D. hispida* oligosaccharides (DHOS).

3.2. *Growth of lactic acid bacteria and bifidobacterium on D. hispida oligosaccharides, inulin and FOS*

The increase in viable count of the microbes in different substrates after 24 h fermentation was recorded in Table 1. It shows that all microbes can grow in the substrates. In fact, *D. hispida* oligosaccharides and commercial prebiotics also supported the growth of pathogens. Thus, increase in viable counts of beneficial bacteria alone cannot be used as indicator to prebiotic characteristic of a substrate.

Table 1. Increase in the cell density of selected lactobacilli, bifidobacterium and pathogens in media containing glucose, *D. hispida* oligosaccharides (DHOS), FOS and inulin after 24 h of fermentation.

Bacterial cultures	Cell density ($\log_{10}$ CFU/mL)*			
	Glucose	DHOS	FOS	Inulin
L. acidophilus FTDC 2131	2.55 ± 0.13^c	2.72 ± 0.14^a	2.69 ± 0.10^b	2.63 ± 0.04^b
L. casei FTCC 0442	1.75 ± 0.01^d	2.24 ± 0.01^c	2.65 ± 0.04^b	2.90 ± 0.12^a
L. casei ATCC 393	3.17 ± 0.90^a	3.15 ± 0.06^a	2.96 ± 0.04^c	3.07 ± 0.10^b
L. plantarum FTCC 0350	2.46 ± 0.12^b	2.52 ± 0.07^a	2.47 ± 0.03^b	2.10 ± 0.08^c
Bifidobacterium bifidum 12	2.33 ± 0.01^d	3.53 ± 0.03^a	3.25 ± 0.03^b	3.10 ± 0.09^c
Escherichia coli FTDC 201	2.14 ± 0.11^c	1.73 ± 0.01^d	2.34 ± 0.02^b	2.82 ± 0.08^a
Campylobacter coli c159	2.24 ± 0.10^a	2.14 ± 0.02^b	2.02 ± 0.01^c	2.16 ± 0.11^b
Clostridium perfringens c438	2.03 ± 0.08^a	1.62 ± 0.01^c	1.82 ± 0.10^b	1.63 ± 0.03^c

* Increase in cell density was the differences between the viable count at 24 h and 0 h (in $\log_{10}$ CFU/mL).
[abcd]Different superscript indicate significant different within a row for different bacteria ($p < 0.05$).

Therefore, Huebner et al. [13] had introduced a formulae to calculate prebiotic activity score, taking into account that a substrate, be it prebiotic or not a prebiotic, could support growth of both beneficial microbes and pathogens.

IOP Conf. Series: Materials Science and Engineering **716** (2020) 012005 doi:10.1088/1757-899X/716/1/012005

Substrate with prebiotic characteristic should selectively support the growth of beneficial microbes more than the pathogens. Prebiotic activity score is calculated by the difference of increase in viable count of beneficial microbes in potential prebiotic over glucose and increase in viable count of pathogens in potential prebiotic over glucose. Positive and big score indicates a good prebiotic substrate. Many researchers interested in studying oligosaccharides from various sources to be claimed as prebiotics such as oligosaccharides from dragon fruit flesh [14], pectic oligosaccharides from orange peel wastes [15], oligosaccharides from coffee mannan [16] and refined arabinoxylooligosaccharides from wheat bran [17]. All of these oligosaccharides have supported the growth of beneficial microorganisms.

3.3. Prebiotic Activity Score

Figures 2(a-c) show the prebiotic activity scores (PAS) of *D. hispida* oligosaccharides and commercial prebiotics for five beneficial microbes; *L. acidophilus* FTDC 2131, *L. casei* FTCC 0442, *L. casei* ATCC 393, *L. plantarum* FTCC 0350 and *B. bifidum* 12 against three detrimental microbes; *E. coli*, *C. coli* and *C. perfringens*. Higher and positive prebiotic score indicates better substrate. Generally, PAS of *D. hispida* oligosaccharides and commercial prebiotics (FOS and inulin) were positive and significantly higher than that of blank medium (medium with glucose). *D. hispida* oligosaccharides had higher PAS than inulin for all the beneficial microbes ($p < 0.05$) except *L. plantarum* against *E. coli*, *L. casei* FTCC 0442 against *C. perfringens* and *C. coli*. *D. hispida* oligosaccharides exhibited high PAS with *L. casei* FTCC 0442 (+0.471) and *B. bifidum* 12 (+0.470) against *E. coli*, with *B. bifidum* 12 against *C. coli* (+0.613) and *C. perfringens* (+0.713). Commercial FOS scored positive PAS for all beneficial microbes against *E. coli* (+0.04 to +0.171), *C. coli* (+0.034 to +0.500) and *C. perfringens* (+0.037 to +0.501). Commercial prebiotic inulin had positive PAS for most beneficial microbes but inulin PAS were negative for *L. acidophilus* FTDC 2131 (-0.287) and *L. casei* ATCC 393 (-0.392) against *E. coli*, and for *L. plantarum* FTCC 0350 against *C. coli* (-0.151). Thus, it was assumed that the prebiotic activity score of each substrate is largely depending on the substrate itself and type of bacterial strains. PAS of longan fruit hydrolysates was +1.69 with *L. acidophilus* La5 and PAS of the same substrate with *B. lactis* was +1.44 [18].

Nevertheless, pairing *D. hispida* oligosaccharides with *B. bifidum* 12 scored the highest PAS and thus the fermentation media of different substrates by *B. bifidum* 12 were further analysed for lactic, acetic and propionic acids.

3.4. Concentration of lactic, acetic and propionic acids

Table 2 displays the concentration of lactic, acetic and propionic acids after 24 h fermentation of substrates by *B. bifidum* 12. Fermentation of *D. hispida* oligosaccharides produced significantly the highest concentration of acetic and propionic acids. The highest lactic acid produced by synergistic of *D. hispida* oligosaccharides and *B. bifidum* 12 was 30.42 mM, significantly lower than FOS but higher than inulin. Ability to produce lactic and short chain fatty acids is one of 8 criteria for a probiotic [19]. Lactic and short fatty acids reduce the intraluminal pH which favours the growth of bifidogenic and other lactic acid bacteria. Short chain fatty acids also stimulate water and electrolyte absorption in the intestine and hence reduce the risk of diarrhoea and dehydration. They also increase colonocyte proliferation and metabolic energy production [20].

IOP Conf. Series: Materials Science and Engineering **716** (2020) 012005 doi:10.1088/1757-899X/716/1/012005

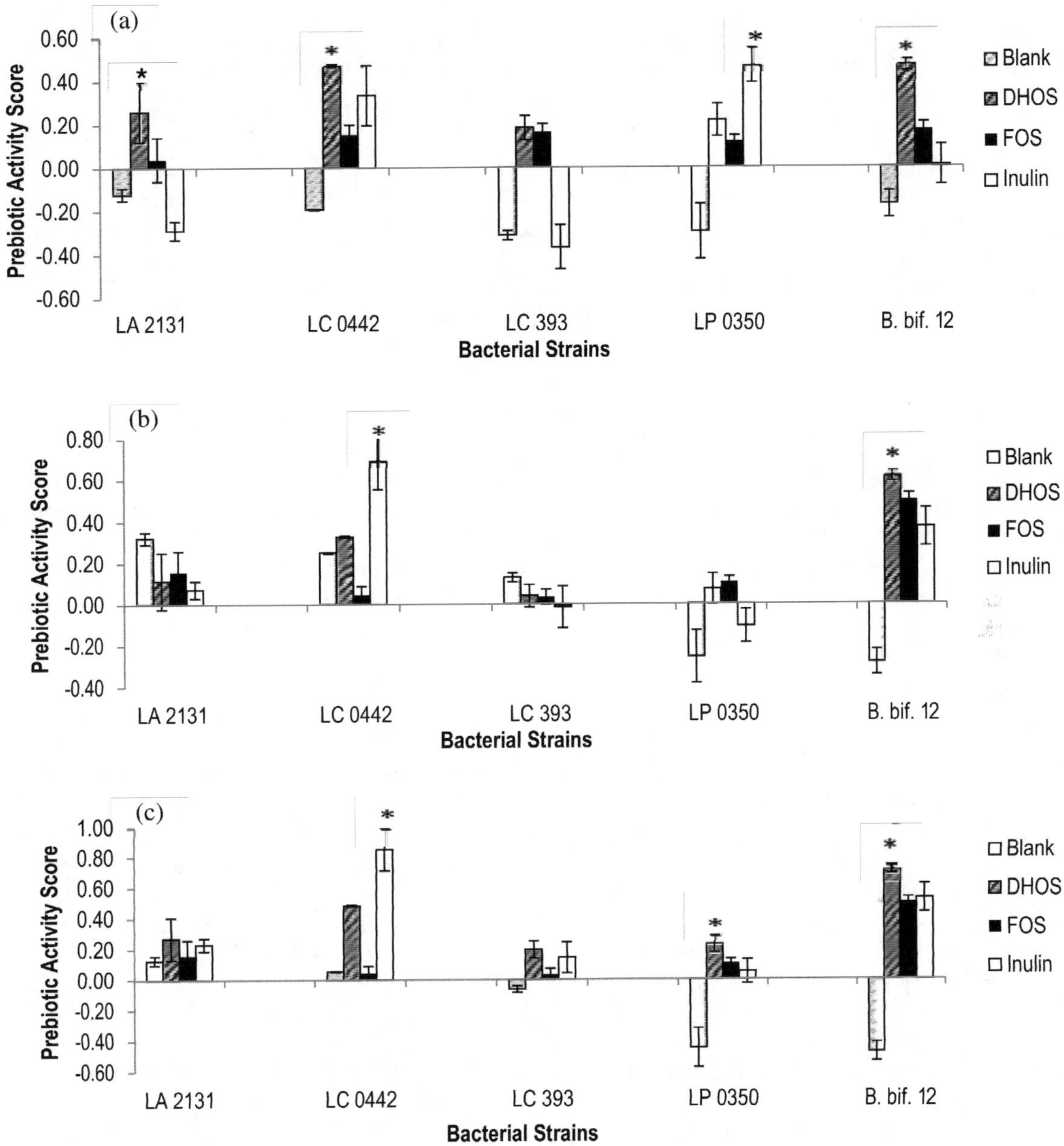

Figure 2. Prebiotic activity scores for of bacterial cultures supplemented with 1% substrates (DHOS, FOS and inulin) by lactobacilli and bifidobacterium against (a) *Escherichia coli*, (b) *Campylobacter coli* and (c) *Clostridium perfringens*. Data were expressed as mean ± SD from calculation of three independent experiments.*Bar indicates the highest score within the same bacterial strains ($p < 0.05$). LA2131, *L. acidophilus* FTDC2131; LC0442, *L. casei* FTCC0442; LC393, *L. casei* ATCC393; LP0350, *L. plantarum* FTCC0350; B. bif 12, *B. bifidum* 12.

Table 2. Concentration of lactic, acetic and propionic acids after 24 h fermentation of different substrates by *B. bifidum* 12.

Substrate	Lactic acid (mM)	Acetic acid (mM)	Propionic acid (mM)
Glucose	53.95 ± 1.72[a]	19.45± 0.01[b]	0.55 ± 0.01[c]
DHOS	30.42 ± 0.33[b]	22.63 ± 0.14[a]	0.71 ± 0.06[a]
FOS	53.92 ± 0.09[a]	13.31 ± 0.11[c]	0.69 ± 0.21[b]
Inulin	15.47 ± 0.88[c]	19.00 ± 0.09[b]	0.63 ± 0.05[b]

Data were expressed as mean ± SD from three independent experiments.
[abc]Different superscript alphabets indicate significant different concentrations of a particular acid among different substrates ($p < 0.05$).

4. Conclusion

D. hispida oligosaccharides were able to selectively enhance the growth of lactobacilli and bifidobacterium strains better than the growth of pathogens. *D. hispida* oligosaccharides was the most preferred substrate by *L. acidophilus* FTDC 2131, *L. plantarum* FTCC 0350 and *B. bifidum* 12 as compared to inulin and FOS. Fermentation of DHOS by *B. bifidum* 12 had significantly exhibited the highest PAS against the pathogens and produced substantial lactic (30.42 mM), acetic (22.63 mM) and propionic (0.71 mM) acids.

References

[1] Agbor-Egbe, T and Treche, S 1995 Evaluation of the chemical composition of Cameroonian yam germplasm *J Food Compos Anal* **8**(3) 274-83.

[2] Tattiyakul, J, Asavasaksakul, S and Pradipasena, P 2006 Chemical and physical properties of flour extracted from taro *Colocasia esculenta* (L.) Schott grown in different regions of Thailand *ScienceAsia* **32**(3) 279-84.

[3] Nashriyah, M, Nornasuha, Y, Salmah, T, Norhayati, N and Rohaizad, M 2010 *Dioscorea hispida* Dennst. (Dioscoreaceae): An Overview *Buletin UniSZA* **4** 2180-235.

[4] Miah, M M, Das, P, Ibrahim, Y, Shajib, M S and Rashid, M A 2018 *In vitro* antioxidant, antimicrobial, membrane stabilization and thrombolytic activities of *Dioscorea hispida* Dennst *Eur J Integr Med* **19**(4) 121-7.

[5] Liu, Q, Donner, E, Yin, Y, Huang, R L and Fan, M Z 2006 The physicochemical properties and in vitro digestibility of selected cereals, tubers and legumes grown in China *Food Chem* **99** 470-7.

[6] Nudin, N F H, Ali, A M, Ngah, N, Mazlan, N Z, Mat, N, Ghani, M N A, Alias, N, Zakaria, A J and Jahan, M S 2017 ISSR marker-assisted genetic diversity analysis of *Dioscorea hispida* and selection of the best variety for sustainable production. *Comptes Rendus Biologies* **340**(8) 359-66.

[7] Ashri, A, Yusof, M S M, Jamil, M S, Abdullah, A, Yusoff, S F M, Arip, M N M and Lazim, A M 2014 Physicochemical characterization of starch extracted from Malaysian wild yam (*Dioscorea hispida* Dennst.) *Emir J Food Agric* **26**(8) 652-8.

[8] Kuo, S M 2013 The interplay between fiber and the intestinal microbiome in the inflammatory response *Adv Nutr* **4**(1) 16–28.

[9] National Center for Complementary and Integrative Health (NCCIH) 2018 Probiotics: In Depth National Institutes of Health, U.S. Department of Health and Human Services, Publication D345 Assessed on 31 May 2019 from https://nccih.nih.gov/health/probiotics/introduction.html.

[10] Pandey, K R, Naik, S R and Vakil, B V 2015 Probiotics, prebiotics and synbiotics-a review *J Food Sci Tech* **52**(12) 7577-87.

[11] Hoover, R and Senanayake, N 1996 Effect of sugars on the thermal and retrogradation properties of oat starches *J Food Biochem* **20**(1) 65-83.

[12] Haska, N and Ohta, Y 1991 Pretreatment effects on the hydrolysis of sago starch granules by raw starch-digesting amylase from *Penicillium brunneum Proc of the 4th Int Sago Symp,* eds T T Ng et al (Sarawak Malaysia: Lee Ming Press) pp 53-63.

[13] Huebner, J, Wehling, R L and Hutkins, R W 2007 Functional activity of commercial prebiotics *Int Dairy J* **17**(7) 770-5.

[14] Wichienchot, S, Jatupornpipat, M and Rastall, R A 2010 Oligosaccharides of pitaya (dragon fruit) flesh and their prebiotic properties *Food Chem* **120**(3) 850-7.

[15] Gomez, B, Gullon, B, Remoroza, C, Schols, H A, Parajo, J C and Alonso, J 2014 Purification, characterization and prebiotic properties of pectic oligosaccharides from orange peel wastes *J Agr Food Chem* **62**(40), 9769-82.

[16] Asano, I, Hamaguchi, K, Fujii, S and Iino, H 2003 *In vitro* digestibility and fermentation of mannooligosaccharides from coffee mannan *Food Sci Technol Res* **9**(1) 62-6.

[17] Gullon, B, Gullon, P, Tavaria, F, Pintado, M, Gomes, A M, Alonso, J L and Parajo, J C 2014 Structural features and assessment of prebiotic activity of refined arabinoxylooligosaccharides from wheat bran *J Funct Food* **6**(1) 438-49.

[18] Thitiratsakul, B and Anprung, P 2014 Prebiotic activity score and bioactive compounds in longan (*Dimocarpus longan* Lour.): Influence of pectinase in enzyme-assisted extraction *J Food Sci Tech* **51**(9) 1947-55.

[19] FAO/WHO 2006 Probiotics in food: Health and nutritional properties and guidelines for evaluation. FAO Food and Nutrition Paper 85. Accessed on 31 May 2019 from http://www.fao.org/tempref/docrep/fao/009/a0512e/a0512e00.pdf

[20] Van den Abbeele, P, Van de Wiele, T, Grootaert, C, Verstraete, W, Gérard, P, Bruneau, A, Possemiers, S 2010 *Arabinoxylans and Inulin Modulate the Luminal and Mucosa-Associated Bacteria In Vitro and In Vivo,* ed J W Van der Kamp et al (Massachusetts: Wageningen Academic Publishers) pp 223-49.

Acknowledgements
This research was funded by the Malaysian Ministry of Higher Education ERGS 304/PTEKIND/6730099 and Universiti Sultan Zainal Abidin (ASTS fellowship for Napisah H).

IOP Conf. Series: Materials Science and Engineering **716** (2020) 012006 doi:10.1088/1757-899X/716/1/012006

Effect of feeding strategy on the protein and fatty acid contents of black soldier fly prepupae (*Hermetia illucens*) for the potential applications as animal feed and promising alternative protein-rich food

L Mohamad[1], F Dina[2], H A Hasan[2], K Sudesh[2], S Baidurah*[1]

[1]School of Industrial Technology, Universiti Sains Malaysia, 11800, Minden, Penang, Malaysia
[2]School of Biological Sciences, Universiti Sains Malaysia, 11800, Minden, Penang, Malaysia

*Email: sitibaidurah@usm.my

Abstract. There is considerable interest in the exploring of an alternative protein source due to ever-growing population in the world. Black soldier fly (BSF), *Hermetia illucens*, can be a promising alternative protein-rich food due to its high protein content as compared to livestock. This study investigated the effect of different feeding strategies on the crude protein of BSF prepupae (BSFP) to further enhance the protein contents. The feeding strategy was done at BSFP stage using two different diets which were a mixture of food waste as a control, and bacterial dried cells. Protein and fatty acid content analyses were done on the freeze-dried BSFP samples harvested on day 4 after feeding in order to determine the crude protein and fatty acid. BSFP fed with lyophilized cells showed enhancement in the nutritional contents, compared to the conventional feeding strategy using food waste, with increased crude protein content by 17%. This study demonstrated that the bacterial dried cells can be utilized as a single cell protein to further increase the protein content in the BSFP body which can be applied for the animal feed and potential human consumption.

1. Introduction

The global food production system is facing challenges due to the ever-growing population in the world. According to the Food and Agriculture Organization of the United Nations (FAO), with the exponential growth of the global population, it is expected to reach the threshold of 9 billion people by 2050 [1]. This will result in a greater need for food, especially for animal protein sources such as cattle, poultry and fish. In addition, the FAO has foreseen that global meat production and consumption will double by 2050 compared to 2000 [2]. Food production supply will continuously decrease and this situation will unable to satisfy the demand of the global population, hence, will lead to the food-security crisis. Livestock farming is deemed unsustainable and resulted in many environmental impacts such as the accumulation of all anthropogenic greenhouse gases (GHG) [3], space competition with human and contribute to higher water footprint than plant-based foods [4].

To mediate this, a sustainable diet consists of alternative protein sources need to be considered. Insect meals may serve as one of the options to achieve more sustainable production due to their potential food source, high feed conversion efficiency compared to traditional livestock, short period of breeding, less space required during breeding, and high protein content [5]. Moreover, insects are considered the best alternative as a partial or complete substitution for fish meal due to their nutritional contents, ease of rearing and high biomass production of larvae or pupae [6]. Table 1 shows the composition of commonly consumed insects as compared to conventional foods [7,8,9]. Overall, most of the insects contain high amount of protein content which is comparatively higher than the conventional food up to 41%.

Table 1. Protein content of conventional foods and commonly consumed insects [5,7,8,9].

Organism	Protein (%)
Conventional foods	
Beef	18.4
Lamb	15.4
Pork	14.6
Chicken	22.0
Fish	18.3
Insects	
Termites	23.3
Caterpillars	38.1
Adult weevils	30.3
House fly pupae	63.1
May beetle larvae	11.1
Adult female ants	7.4
Adult male ants	25.2
Bee	18.1
Silkworm	23.1
Grasshopper	46.1
Black soldier fly	40.2

Insects such as *Hermetia illucens*, commonly known as black soldier fly (BSF) can be a promising alternative protein-rich food due to its high protein content as compared to livestock [7]. Black soldier fly larvae (BSFL) is regarded as a great potential in the feed sector whereby insect-meal could substitute in part fish-meal and soybean-meal as a protein source in the animal and aquaculture feed industry [8]. The BSFL can consume a wide range of biowaste, such as fruits and vegetable waste, kitchen waste, abattoir waste and animal manure, which will directly affect the nutritional value of the black soldier fly prepupae (BSFP) [9].

The aim of this study is to further increase the protein content of BSFP for the applications of animal feed and promising alternative protein-rich food. The feeding strategy was modified by feeding the BSFP with bacterial dried cells instead of conventional feeding with food waste, with the aim to increase the crude protein content of BSFP. The obtained BSFP was further subjected to the proximate analysis to determine the protein and lipid content. In detail, the feeding strategy was done at BSFP stage using two different diets which were a mixture of food waste as a control, and bacterial dried cells containing intracellular polyhydroxybutyrate (PHB), a type of biodegradable polymer. *Cupriavidus necator* wild type H16 (ATCC 17699), formerly known as *Ralstonia eutropha* was the strain used in this modified feeding strategy. This strain is capable of accumulating intracellular granules of PHB under growth media rich with carbon source and limiting nitrogen source [10]. PHB and its copolymers are biodegradable polymers that have similar properties as conventional petrochemical plastics with nontoxic and biocompatibility behaviour which are of particular interest [10]. In the latter feeding strategy, two products were obtained: (1) the BSFP as a source for animal feed and as a promising alternative protein-rich food, and (2) the extracted PHB through biological extraction method, by utilizing the intestine of BSFP.

2. MATERIALS AND METHODS

2.1 Organism and culture media

Cupriavidus necator wild type H16 (ATCC 17699) was used throughout this study. The bacterial cell was sub-cultured on the agar plate for short-term storage. For long-term preservation, the bacterial cells were maintained in a 40% (v/v) glycerol stock solution. The strain was sub-cultured on the nutrient-rich (NR) agar medium consisting per liter: 2 g yeast extract, 10 g peptone, 10 g meat extract and 15 g of bacteriological agar powder [11]. The pH of the medium was adjusted to 7 by using 1 M of hydrochloric acid (HCl) and 1 M of sodium hydroxide (NaOH). The medium was sterilized by autoclaving at 121 °C for 15 minutes using the autoclave machine (Model: TOMY-ES-315, Japan). NR broth was prepared using the same method as the NR agar medium, but without adding bacteriological agar powder.

2.2 Bacterial cultivation

Bacterial cultivation used in this study was basically the same as that described in our previous paper with a slight modification [10]. *C. necator* bacterial colony on the NR agar plate was inoculated into NR broth and grown in the incubator shaker (Model: Infors Ecotron, New Zealand) at 200 rpm at 30 °C, 24 hours for cell activation. After the cell activation, 15 mL of the NR culture was inoculated into 300 mL of mineral medium (MM) in 500 mL shake flask. The fermentation media consists of MM and minor element solutions with fructose concentration of 20 g/L as the sole carbon source. The MM composed of 4.6 g/L Na_2HPO_4, 4.0 g/L NaH_2PO_4, 0.39 g/L $MgSO_4$, 62 mg/L $CaCl_2$, 0.45 g/L K_2SO_4, and 1 mL/L of trace element solution. The trace element solution was prepared by dissolving 2.4 g/L $MnSO_4 \cdot H_2O$, 0.48 g/L $CuSO_4 \cdot 5H_2O$, 15 g/L $FeSO_4 \cdot 7H_2O$, and 2.4 g/L $ZnSO_4 \cdot 7H_2O$, in 1 L of 0.1 M HCl [11]. The culture was incubated in an incubator shaker at 200 rpm at 30 °C, 48 hours for production of polyhydroxybutyrate (PHB). After 48 hours of fermentation, the culture was centrifuged at 8590 *g* and 4 °C for 8 minutes using bench top centrifuge (Model: KUBOTA 5100, Japan). The cell pellets collected were washed with distilled water and kept at -20 °C prior to freeze-drying process [11].

2.3 Cell freeze-drying

The frozen cell pellets were freeze-dried using freeze dryer machine (Model: LABCONCO, US) at temperature -50 °C with pressure range from 0.06 to 0.09 Torr for 48 hours. The obtained freeze-dried bacterial cells were subjected to feeding experiment.

2.4 Feeding strategy of black soldier flies prepupae

The BSF were obtained from School of Biological Sciences, Universiti Sains Malaysia (USM) and placed in plastic containers (20 x 10 x 18 cm) covered with a net. BSFL were fed with the conventional diet consisting of food waste collected from Bakti Cafeteria, USM for the survival and larval development. Once it turns into pre-pupae, the BSFP will crawl out of the food waste container to find a hiding place. A total of 400 BSFP was collected and starved for 48 hours in a different container containing sawdust. Prior to the feeding experiment, the BSFP were transferred into a clean plastic container and was fed with the freeze-dried *C. necator* once a day for a consecutive 3 days with a feeding rate of 5% of the total body weight per day as described by Murugan et al. [12], utilizing mealworm as a model organism. The BSFP were harvested on day 4 and washed with distilled water to remove the remaining food residue and fecal pellets. Next, the BSFP were placed on tissue paper for the purpose of drying. The BSFP bodies were kept in - 20 °C overnight and were then freeze-dried using the same setting as in section 2.3 for 4 days. The freeze-dried BSFP sample was kept for further proximate analysis.

2.5 Analysis of BSFP

The analysis of BSFP comprised of analytical determination of crude fat, crude protein, crude fiber, ash, moisture and dry matter. The initial analysis procedure is to determine the total lipid content by using the Soxhlet extraction method [13]. The freeze-dried and ground BSFP sample was continuously extracted with petroleum ether for 6 hours and the remaining defatted residue was dried and weighed. The defatted sample was further used for determining crude protein content while the extracted oil sample was further subjected to Gas Chromatography-Mass Spectrophotometry (GC-MS) analysis for fatty acid methyl ester (FAME) determination. Nitrogen determination of BSFP was conducted using the Kjeldahl method [13] which involves digestion, distillation and titration. Protein extraction was done prior to Kjeldahl analysis by using an alkaline extraction method by utilizing NaOH. Detailed methodology is described in the following subsection. The crude protein content was calculated using a protein-to-nitrogen conversion factor of 6.25 (based on protein containing an average of 16% nitrogen) [14].

Crude fiber analysis was performed by boiling the dried defatted sample with 5% H_2SO_4, followed by alkaline hydrolysis with 25% NaOH, which removes some hemicellulose and lignin [13]. The remaining residue after digestion was oven dried and placed in a muffle furnace at 450 °C and burned until no black spot were observed [13]. Crude fiber was measured as the weight loss on incineration of the residue remaining after digestion of the sample. Crude ash, which comprised of non-combustible minerals was analyzed by heating the ground BSFP in a muffle furnace for 4 hours at 550 °C [13]. The moisture content and dry matter were determined by oven-dry the ground freeze-dried BSFP in an air circulating oven at 105 °C until constant weight was achieved [13].

2.6 Protein Extraction

The ether-defatted BSFP product was used for protein extraction by utilizing NaOH [14]. One g of each defatted BSFP powdered sample was weighed accurately into separate triplicate falcon tube. The powdered samples were then dissolved in 15 ml of 0.25 M NaOH [14]. In order to enhance the extraction process, this steps was carried out in an incubator shaker set at 40 °C for 1 hour at 100 rpm. Next, the suspension was centrifuged at 4 °C for 20 minutes at 3500 *g* [14]. The second extraction was performed on the remaining suspended pellet collected at the base of the tube. The pH of the supernatant and gel layer collected from both extractions were adjusted to 4.3 - 4.5 with 2 M HCl at room temperature [14]. The precipitated protein was then harvested by centrifugation at 4 °C for 15 minutes at 2500 *g* [14]. The supernatant and gel layer were discarded while the pellet containing the desired protein precipitate was washed 3 times with distilled water to rinse out HCl traces followed by centrifugation at 4 °C for 10 minutes and 2500 *g* [14]. The precipitate was frozen at -20 °C overnight (about 18 hours). The final protein extract with moisture content less than 5 % was obtained by freeze-dried the frozen precipitate.

2.7 Preparation of fatty acid methyl ester (FAME) derived from BSF

Fatty acid analysis consisted of two consecutive steps, preparation of fatty acid methyl ester (FAME) and chromatographic analysis [15]. FAME was synthesized from the lipid sample by heating 0.3 g of the oil sample with 5 mL of methanolic NaOH for esterification. The sample was boiled for 5 to 10 minutes under reflux until the fat droplets disappear. The sample was esterified, and 7 mL of the methanolic boron trifluoride solution was added and boiled for another 2 minutes. The amount of 3 mL of *n*-heptane was added and boiled for 1 minute in order to recover the methyl ester in an organic phase. The mixture was cooled to room temperature and followed by the phase separation of the aqueous and organic layers by adding saturated NaCl solution in the mixture. The upper layer of *n*-heptane phase was pipetted out into the test tube, and a small amount of anhydrous Na_2SO_4 was added to remove the traces of water from organic solutions. The methyl ester solution was transferred into vials and subjected to GC-MS analysis to determine the fatty acids composition.

2.8 Analysis of fatty acid methyl ester (FAME) using gas chromatography-mass spectroscopy (GC-MS)

IOP Conf. Series: Materials Science and Engineering **716** (2020) 012006 doi:10.1088/1757-899X/716/1/012006

Fatty acid methyl ester (FAME) analysis was performed using GCMS-QP2010 Ultra Gas Chromatograph Mass Spectrometer (Shimadzu) and equipped with SUPELCOWAX® 10 Capillary GC Column (30 m × 0.25 mm i.d; 0.25 µm film thickness). The oven temperature was set at 115 °C, raised to 160 °C at a rate of 10 °C/min, held for 2 min, and finally raised to 225 °C at a rate of 10 °C/min where it was held for another 12 min. The injection volume was 0.2 µL using helium as the carrier gas at a rate of 1.2 mL/min with a split ratio of 50:1 at the inlet. Identification of methyl esters was made by comparing the mass spectra with the mass spectral database.

3. RESULTS AND DISCUSSION

The comparison of nutrient compositions of BSFP is shown in table 2. The crude protein content of freeze-dried BSFP fed with bacterial-dried cells was comparably higher with a value of 63.67%, which was the highest as compared to studies reported by Ruhnke et al. [19], Liu et al. [5] and Spranghers et al. [17]. This observation indicated that feeding strategy using lyophilized cells which acts as a single cell protein (SCP) contributes to the higher protein content inside the BSFP body. The SCP refers to protein derived from the cells of microorganisms which are grown on various carbon sources such as bacteria, fungi, algae and yeast [16]. The protein content of *C. necator* can reach up to 74% of the total cellular content compared to yeast, fungi and algae which was around 50% [16]. Furthermore, a study carried out by Murugan et al. [12] on the experimental feeding of mealworms with *C. necator* showed a significant increase of protein in the body compared to mealworms fed with oats (conventional diet) which was parallel with the findings in this study by utilizing BSF as a model organism fed with *C. necator* showed increase in protein content up to 17%.

Table 2. Comparison of nutrient compositional analyses of black soldier fly (BSF) larvae (% in dry matter).

No.	Nutrient content (%)	This study	References		
			Feeding with organic food waste		
		Feeding with bacterial dried cells	[19]	[5]	[17]
1.	Crude Protein	63.67 *	46.70	40.20	43.10
2.	Crude Fat	24.30	42.20	28.00	38.60
3.	Crude Fibre	4.22	1.56	-	-
4.	Ash	9.90	5.56	8.80	2.70
5.	Moisture Content	4.23	-	-	-
6.	Dry Matter	95.77	-	-	-

* Relative standard deviation = 8.44 % (n = 3)

BSF is an edible insect which contains high amount of protein and fat. The BSF protein composition can be manipulated by using different substrates in a rearing conditions [17]. The BSFL used in this study were fed with a mixture of food waste for survival and development, which latter fed with bacterial dried cells at the stage of early prepupae. In this study, the BSFP were fed with lyophilized cells of *C. necator* cultivated on fructose as a sole of carbon source. The BSFP were fed daily for a period of 3 days with 3.16 g of freeze-dried bacterial cells, which is about 5 wt% of the total BSFP body weight. In a recent report on the biological recovery approach for PHB using mealworm by Murugan et al. [12], on average, mealworm consumed about 5 wt% freeze-dried bacterial cells of their total body weight daily. Therefore, this feeding rate can be applied to BSFP since the basic structure of the internal guts is similar across insects, although they possess a diversity of modification associated with adaptation to different feeding states [18].

The crude fat contents of BSF obtained in this study were rather low compared to the crude fat levels reported by Ruhnke et al. [19], Liu et al. [5], and Spranghers et al. [17]. BSFL reared on restaurant waste

[17] had slightly higher crude fat content (38.6%) compared to this current study (24.3%). For both experimental diets, the BSFL were fed with kitchen waste during the larvae development and the nutrient was incorporated into their bodies. The food waste used in this study contained a mixture of protein, fat, carbohydrate and fibre. The food waste collected were segregate according to their nutritional content, for example chicken, meat or fish waste for protein, excess oil accumulated from the food waste for fat, rice waste for carbohydrate and vegetables or fruit pulp waste for fibre. By comparison, the nutritional content in the food waste obtained might be varied at a different location. Hence, this will directly affect the accumulation of nutrient as well as the lipid content inside the BSFP body. In addition, BSFL reared on brewery spent grain [19] had the highest crude fat content (42.2%) compared to those reared on other substrates. Generally, a decreasing trend in fat content is observed with an increase in the protein content. In terms of application of BSF for animal feed, high levels of lipids in the fish diet can lead to problems such as excessive fat deposition in the liver, which may decrease the fish health, quality and shelf life of the final product [20]. Hence, our study demonstrated a favorable property to be applied in animal feed industries.

The effects of the feeding strategy on the FAME composition derived from the BSFP biomass was determined and the obtained results for GC-MS analysis of BSFP FAME was illustrated in figure 1. The first peak representing capric acid methyl ester (C10:0) (1.57%) appeared after 2.487 minutes while the other major peaks, representing lauric acid methyl ester (C12:0) (48.70%) appeared at 3.702 minutes, myristic acid methyl ester (C14:0) (14.07%) at 5.334 minutes, palmitic acid methyl ester (C16:0) (9.78%) at 7.166 minutes, palmitoleic acid methyl ester (C16:1) (4.17%) at 7.418 minutes, stearic acid methyl ester (C18:0) (1.19%) at 8.969 minutes, oleic acid methyl ester (C18:1) (13.49%) at 9.205 minutes and linoleic acid methyl ester (C18:2) (5.17%) at 9.672 minutes.

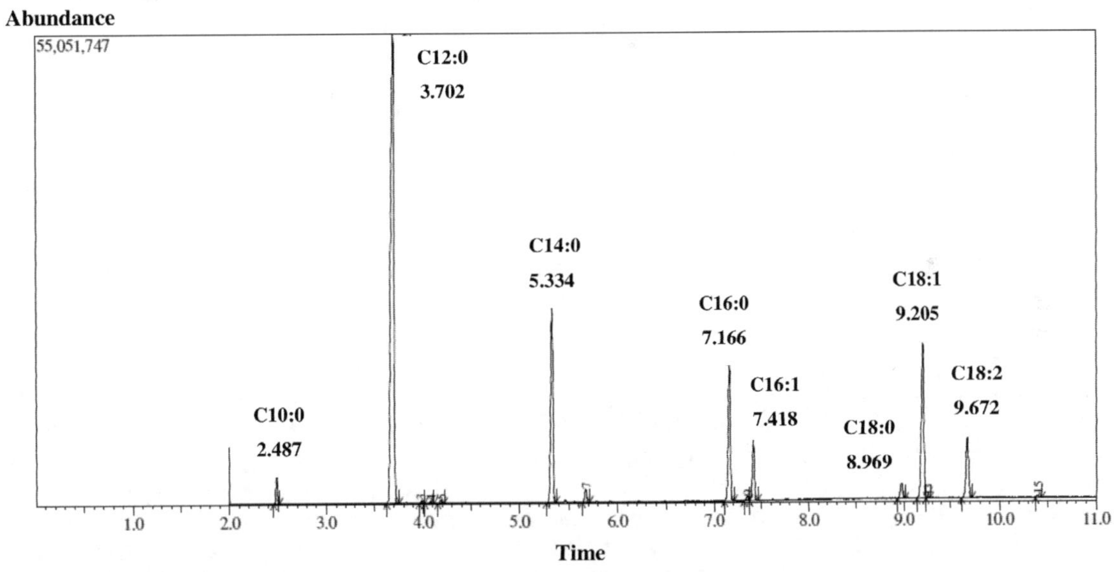

Figure 1. GC-MS analysis results of BSFP FAME.

Referring to table 3, the FAME of BSFP content are comparatively the same with other reported study, with little changes in the relative content. The BSFP FAME composition in this study was found to contain higher saturated fatty acid as compared to unsaturated fatty acid composition, which were in reasonable agreement with the results reported by [21]. In detail, the highest composition of FAME produced by BSFP after three times feeding in this study was lauric acid methyl ester (C12:0) with 48.70%. This notable trend of fatty acid composition corresponds to the previous study [21], despite undergoing different feeding strategy, the highest composition of lauric acid (27.80%) was observed. Overall, the results revealed that even though the current study produces lower total crude fat content;

24.30% as compared to 42.20% [19], the compositions of FAME synthesized from BSFP fed with bacterial dried cells were parallel to the fatty acid profile from other reported finding [21].

Table 3. FAME composition (%) derived from BSF fed with various feeding strategies

FAME Composition	Value (%)		
	This study		[21]
	Control normal feeding with food waste	3 times feeding with bacterial cell	Feeding with pre-consumer waste
Capric acid methyl ester (C10:0)	1.02	1.57	0.6
Lauric acid methyl ester (C12:0)	36.96	48.70	27.8
Myristic acid methyl ester (C14:0)	7.98	14.07	5.9
Palmitic acid methyl ester (C16:0)	15.30	9.78	20.4
Palmitoleic acid methyl ester (C16:1)	2.56	4.17	4.0
Stearic acid methyl ester (C18:0)	1.59	1.19	3.0
Oleic acid methyl ester (C18:1)	20.62	13.49	24.8
Linoleic acid methyl ester (C18:2)	10.46	5.17	13.2
Linolenic acid methyl ester (18:3)	0.43	Nil	Nil

Nil = zero

4. CONCLUSION

The issue of the increasing global demand for protein in future can be prevented by replacing the conventional protein-based food with BSFP as an alternative protein source. This entomophagy, however, is still new not only in Malaysia but also in most parts of the world. Protein derived from this edible insect can be extracted using an alkaline extraction method by utilizing NaOH. In this study, by feeding the BSFP with bacterial dried cells have successfully increased the protein content up to 17% as compared to the previous reported studies of BSFL reared on brewery spent grain by Ruhnke et al. [19]. This feeding strategy has also proved that BSF could digest the lyophilized cells of *C. necator* and the protein content in the BSF body was enriched. Furthermore, this feeding strategy has successfully extract the intracellular PHB of *C. necator* in a fecal pellets forms by utilizing the guts of BSFL. The purity of biologically extracted PHB granules washed with distilled water is 82.95%.

IOP Conf. Series: Materials Science and Engineering **716** (2020) 012006 doi:10.1088/1757-899X/716/1/012006

REFERENCES

[1] FAO 2018 The future of food and agriculture – Alternative pathways to 2050 Rome 224 pp Licence: CC BY-NC-SA 3.0 IGO.

[2] Miglietta P P, De Leo F, Ruberti M and Massari S 2015 Mealworms for food: A water footprint perspective *Water* **7**(11) 6190-6203

[3] Gerber, P J, Steinfeld H, Henderson B, Mottet A, Opio C, Dijkman J, Falcucci A and Tempio G 2013 Tackling climate change through livestock – A global assessment of emissions and mitigation opportunities FAO, Rome

[4] Mekonnen M M and Hoekstra A Y 2012 A global assessment of the water footprint of farm animal products *Ecosystems* **15**(3) 401–415

[5] Liu X, Chen X, Wang H, Yang Q, Ur Rehman K, Li W, Cai M, Li Q, Mazza L, Zhang J, et al. 2017 Dynamic changes of nutrient composition throughout the entire life cycle of black soldier fly *PLoS ONE* **12**(8) 1–21

[6] Tran G, Heuzé V and Makkar H P S 2015 Insects in fish diets *Anim. Front.* **5**(2) 37–44

[7] Tschirner M and Simon S 2015 Influence of different growing substrates and processing on the nutrient composition of black soldier fly larve destines for animal feed *J. Insects Food and Feed* **1**(4) 249–259

[8] Boccazzi IV, Ottoboni M, Martin E, Comandatore F, Vallone L, Spranghers T, Eeckhout M, Mereghetti V, Pinotti L and Epis S 2017 A survey of the mycobiota associated with larvae of the black soldier fly (*Hermetia illucens*) reared for feed production *PLoS ONE* **12**(8) 1–15

[9] Nguyen T T X, Tomberlin J K and Vanlaerhoven S 2015 Ability of black soldier fly (Diptera: Stratiomyidae) larvae to recycle food waste *Environ. Entomol.* **44**(2) 406–410

[10] Baidurah S, Kubo Y, Ishida Y and Yamane T 2018 Direct determination of poly(3 hydroxybutyrate) accumulated in bacteria by thermally assisted hydrolysis and methylation gas chromatography in the presence of organic alkali *Pure Appl. Chem.* **90**(6) 1011–1017

[11] Murugan P, Gan C Y and Sudesh K 2017 Biosynthesis of P(3HB-*co*-3HHx) with improved molecular weights from a mixture of palm olein and fructose by *Cupriavidus necator* Re2058/pCB113 *Int. J. of Biol. Macromolecules* **102** 1112–1119

[12] Murugan P, Han L, Gan C Y, Maurer F H J and Sudesh K 2016 A new biological recovery approach for PHA using mealworm *Tenebrio molitor J. Biotechnol.* **239** 98–105

[13] Horwitz W and Latimer G 2005 Official methods of analysis AOAC International (18th Ed.) Gaithersburg, MD, USA

[14] Zhao X, Vázquez-Gutiérrez J L, Johansson D P, Landberg R and Langton M 2016 Yellow mealworm protein for food purposes - Extraction and functional properties *PLoS ONE* **11**(2) 1–17

[15] Indarti E, Majid M I A, Hashim R and Chong A 2005 Direct FAME synthesis for rapid total lipid analysis from fish oil and cod liver oil *J. Food Compos. Anal.* **18**(2–3) 161–170

[16] Chee J Y, Lakshmanan M, Jeepery I F, Hairudin N H M and Sudesh K 2019 The potential application of *Cupriavidus necator* as polyhydroxyalkanoates producer and single cell protein: A review on scientific, cultural and religious perspectives *Appl. Food Biotechnol.* **6**(1) 19–34

[17] Spranghers T, Ottoboni M, Klootwijk C, Ovyn A, Deboosere S, De Meulenaer B, Michiels J, Eeckhout M, De Clercq P and De Smet S 2017 Nutritional composition of black soldier fly (*Hermetia illucens*) prepupae reared on different organic waste substrates *J. Sci. Food Agric.* **97** 2594–2600

[18] Engel P and Moran N A 2013 The gut microbiota of insects-diversity in structure and function *FEMS Microbiol. Rev.* **37** 699-735

[19] Ruhnke I, Normant C, Campbell D L M, Iqbal Z, Lee C, Hinch G N and Roberts J 2018 Impact of on-range choice feeding with black soldier fly larvae (*Hermetia illucens*) on flock performance, egg quality and range use of free-range laying hens *Anim. Nutr.* **4**(4) 452–460

[20] Craig S, Helfrich L A 2002 Understanding fish nutrition, feeds and feeding, Virginia Cooperative Extension **63** 256-270

[21] Hong T, Chandiramani N, Restrepo-Cano J and Sarathy S M 2018 New approach to improving fuel and combustion characteristics of black soldier fly oil *Chem. Eng. Trans.* **65** 31–36

ACKNOWLEDGEMENTS

The authors greatly acknowledge RUI Grant, Universiti Sains Malaysia (1001/PTEKIND/8011022) for the financial support. Lydia Mohamad expresses her in-depth gratitude towards Universiti Sains Malaysia for supporting her MSc studies through Kuok Foundation MSM Fellowship Award.

IOP Conf. Series: Materials Science and Engineering **716** (2020) 012007 doi:10.1088/1757-899X/716/1/012007

Inhibition of Carbohydrate Hydrolysing Enzymes, Antioxidant Activity and Polyphenolic Content of *Beilschmiedia* Species Extracts

Y N Tay[1], M H A Bakar[1*], M N Azmi[2], N A Saad[2], K Awang[3], M Litaudon[4], and M A Kassim[1]

[1] Bioprocess Technology Division, School of Industrial Technology, Universiti Sains Malaysia, 11800, Minden, Gelugor, Penang, Malaysia

[2] School of Chemical Sciences, Universiti Sains Malaysia, 11800, Minden, Gelugor, Penang, Malaysia

[3] Centre for Natural Product Research and Drug Discovery (CENAR), Department of Chemistry, Faculty of Science, University of Malaya, 50603 Kuala Lumpur, Malaysia

[4] Institut de Chimie des Substances Naturelles, CNRS-ICSN UPR2301, Univ. Paris-Sud 11, av. de la Terrasse, 91198 Gif-sur-Yvette, France

*Email: mhafizi88@usm.my

Abstract. The goal of the present study was to provide *in vitro* evidence for potential inhibition of carbohydrate hydrolysing enzymes and antioxidant activities of methanol and ethyl acetate extracts from barks of two different *Beilschmiedia* species. These extracts were tested on α-amylase and α-glucosidase inhibitory activities, mode of enzyme inhibition, total polyphenolic content (TPC) and antioxidant capabilities. Methanolic bark extract of *Beilschmiedia insignis* demonstrated optimum inhibitory effects against α-amylase and α-glucosidase with IC_{50} values of 3.233 μg/mL and 12.357 μg/mL, respectively. Further analysis of inhibition mode revealed that the extract demonstrated a mixed inhibition against both enzymes. In comparison to other extracts, methanolic bark extract of *Beilschmiedia insignis* demonstrated the highest TPC content of 420.393 mg GAE/g extract, lowest IC_{50} value of 12.103 μg/mL for DPPH radical scavenging ability and highest FRAP value of 1904.247 μM Fe (II)/mg extracts, indicating the antioxidant potential of the extract. A significant strong correlation coefficient was observed between TPC with FRAP ($r = 0.994$, $p < 0.01$) and TPC with DPPH ($r = -0.860$, $p < 0.01$), signifying that antioxidant activity and reducing capability were contributed by the polyphenolic compounds present in the crude extract. Collectively, methanolic bark extract of *Beilschmiedia insignis* possessed significant carbohydrate hydrolyzing enzyme inhibitory effects and antioxidant activity, suggesting its possible alternative application for diabetes and postprandial hyperglycemia treatment.

1. Introduction

In 2012, it was estimated that 1.5 million deaths worldwide were caused by diabetes mellitus (DM), causing it to be ranked number eight among the leading cause of death among both sexes and the fifth leading cause of death in women [1]. DM is characterized by high glucose level in the blood (hyperglycemia) associated by disturbances in carbohydrates, proteins and fats metabolism. Such defects in these metabolic pathways may lead to complete or partial insufficiency of insulin actionand/ or insulin resistance with macro-microvascular complications [2]. Previous studies reported that prolonged hyperglycemia is the key to these DM complications as it causes an increase in oxidative stress due to excess generation of free radicals through protein glycation, glucose oxidation and lipid peroxidation [3]. Therefore, the therapeutic approach for DM should be based on targeting glucose metabolism and oxidative stress caused by hyperglycemia-related conditions [3].

There are numerous conventional drugs available for DM, which varies in mechanism of treatment [4]. One of the pharmacological approaches is by using carbohydrate enzyme inhibitor drugs such as acarbose, voglibose and miglitol [5]. These drugs inhibit both α-amylase and α-glucosidase, which are enzymes responsible for the breakdown of carbohydrates. Such inhibition can be achieved by delaying the carbohydrate digestion, which decrease the rate of glucose absorption, thus maintaining blood glucose level as close to normal as possible. However, these current antidiabetic drugs suffer from a number of undesirable side effects, leading researchers to seek traditional medicinal plants as alternatives for diabetic treatment [4].

Beilschmiedia is a genus of approximately 250 species that are trees or shrubs in the family *Lauraceae* [6]. One of the previous phytochemical investigations of plants of the genus *Beilschmiedia* reported the presence of polyphenols which are the secondary metabolites of plants [7,8]. Polyphenols are organic chemicals characterized by the presence of large multiple phenol structural units. They are among the naturally occurring antidiabetic agents because polyphenols exhibit various important biochemical properties including carbohydrate hydrolysing enzyme inhibition and their capabilities to act as antioxidants [9]. In the present study, the methanol and ethyl acetate extracts from barks of *Beilschmiedia lumutensis* and *Beilschmiedia insignis* were evaluated on carbohydrate hydrolysing enzyme inhibition potential, polyphenolic content and antioxidant activities.

2. Materials and method

2.1. Plant material and preparation crude extracts

The bark of *Beilschmiedia lumutensis* and *Beilschmiedia insignis* were collected at Machang, Kelantan on 2006 and at Kuala Lipis, Pahang on 2007, respectively. This plant was identified by T. Leong Eng, a botanist in University of Malaya. The voucher specimens (KL-5269 and KL-5368, respectively) have been deposited at the Herbarium of the Department of Chemistry, Faculty of Science, University of Malaya, Kuala Lumpur, Malaysia. The air-dried barks of *Beilschmiedia lumutensis* and *Beilschmiedia insignis* (each 200 g) were sliced, ground and extracted with EtOAc (3 × 0.5 L) followed by MeOH (3 × 0.5 L) at room temperature using maceration method. The solvents were filtered and removed using rotary evaporator to gave the respective extracts. Collectively, four crude extracts were produced for biological evaluation in this study, BLM: *Beilschmiedia lumutensis*-methanol; BLEA: *Beilschmiedia lumutensis*-ethyl acetate; BIM: *Beilschmiedia insignis*-methanol; BIEA: *Beilschmiedia insignis*-ethyl acetate. The extracts were carried out for carbohydrate hydrolysing enzyme inhibition potential, polyphenolic content and antioxidant activities.

2.2. α-amylase inhibition assay

The α-amylase inhibition assay was determined according to Kazeem *et al.*, [10] with minor modifications. Briefly, an amount of 250 μL crude extracts of different concentrations (0.004-0.333 mg/mL) was mixed with 250 μL of 0.5 mg/mL α-amylase solution prepared using 0.02 M sodium phosphate buffer (pH 6.9). The mixture was pre-incubated for 10 minutes at 25°C in a water bath. Then, 250 μL of 1% starch solution in 0.02 M sodium phosphate buffer (pH 6.9) was added and further incubated for 10 minutes at 25°C. The reaction was terminated by adding 500 μL of DNS reagent, followed by incubation in boiling water for 5 minutes. The reaction mixture was diluted with 5 mL of distilled water and the absorbance was measured at 540 nm using the Hitachi U-1900 spectrophotometer. Acarbose was used as a positive control. The percentage of enzyme inhibition was calculated by using the following equation:

$$\text{Inhibition (\%)} = \frac{Abs(control) - Abs(sample)}{Abs(control)} \times 100\% \quad (1)$$

IC_{50} which is defined as the concentration of extracts required to achieve 50% of enzyme inhibition was calculated using GraphPad Prism.

2.3. α-glucosidase inhibition assay

The *α*-glucosidase inhibition assay was performed accordingly to a previous method with slight modifications [11]. Yeast α-glucosidase (1U/mL) and pNPG (3 mM) were prepared in 0.02 M phosphate buffer. An amount of 100 µL of *α*-glucosidase enzyme (1U/mL) was added with 10 µL of plant extracts with different concentrations ranging from 0.004 mg/mL to 0.333 mg/mL and incubated for 10 minutes at 37 °C in a water bath. Then, 50 µL of pNPG was added to start the reaction and further incubated at 37 °C for 20 minutes. The reaction was terminated by adding 2 mL 0.1 M sodium carbonate. The absorbance reading of the yellow *p*-nitrophenol was measured at 405 nm using the Hitachi U-1900 spectrophotometer. Acarbose was used as the positive control. Phosphate buffer was used to replace the enzyme for the blank and for negative control, plant extracts were replaced by phosphate buffer. The percentage of enzyme inhibition for *α*-glucosidase was calculated using equation (1) and IC_{50} was calculated using GraphPad Prism.

2.4. Kinetics for mode of α-amylase inhibition

The extract with the lowest IC_{50} value from α-amylase inhibition assay was evaluated for mode of inhibition [10]. Firstly, 250 µL of the extract was mixed with 250 µL of *α*-amylase enzyme. For control group, instead of using the extracts, 250 µL of pH 6.9 phosphate buffer was used. Both mixtures were pre-incubated for 10 minutes at 25 °C in a water bath. Then, 250 µL of starch at increasing concentrations ranging from 0.30 to 5.0 mg/mL was added to both set of the reaction mixtures to initiate the reaction. The mixture was further incubated for 10 minutes at 25 °C and 500 µL of DNS reagent was added to terminate the reaction. The amount of reducing sugars released was determined spectrophotometrically at 540 nm using a maltose standard curve generated using the maltose assay by DNS method. The mode of inhibition on α-amylase was determined by analyzing the Lineweaver-Burk plot using Michaelis-Menten kinetics.

2.5 Kinetics for mode of α-glucosidase inhibition

The extract with the lowest IC_{50} value from *α*-glucosidase inhibition assay was evaluated for mode of inhibition [10]. In brief, 10 µL of the extract was prepared and mixed with 100 µL of *α*-glucosidase enzyme. The mixture was pre-incubated for 10 minutes at 37 °C in a water bath while pH 6.9 phosphate buffer was used to replace the extracts for control group. Then, 50 µL of pNPG dissolved in 20 mM phosphate buffer (pH 6.9) was added at increasing concentrations ranging from 0.63 to 2.0 mg/mL to both set of the reaction mixtures to initiate the reaction. The mixture was further incubated for 10 minutes at 37 °C and 500 µL of sodium carbonate was addedto stop the reaction. The amount of reducing sugars released was determined spectrophotometrically at 405 nm by using a *p*-nitrophenol standard curve. The mode of inhibition of on α-glucosidase was determined by analyzing the Lineweaver-Burk plot using Michaelis-Menten kinetics.

IOP Conf. Series: Materials Science and Engineering **716** (2020) 012007 doi:10.1088/1757-899X/716/1/012007

2.6. Determination of total polyphenolic content (TPC)

The amount of total polyphenolic content (TPC) was determined by using the Folin–Ciocalteu reagent with slight modifications [12]. Each of the methanol and ethyl acetate *Beilschmiedia* extracts was dissolved in DMSO to a final concentration of 0.2 mg/mL. Then, 0.5 mL of the plant extract was dissolved in 2.5 mL of the Folin- Ciocalteu reagent (1/10 dilution) and the mixture was topped up to 10 mL with distilled water. The mixture was left for 5 minutes at room temperature, followed by the addition of 2 mL of 7.5% Na_2CO_3 solution. The final mixture was vortexed and incubated for 2 hours in a dark room at room temperature. Absorbance of the samples were measured at 765 nm using the Hitachi U-1900 spectrophotometer. A gallic acid standard curve was prepared ranging from gallic acid concentrations of 0.2-1.0 mg/mL and TPC was expressed in terms of mg of gallic acid equivalents (GAE) per gram of extract (mg GAE/g extract). The equation below was used for the calculation:

$$\text{TPC (mg GAE/g extract)} = \text{C X} \frac{V}{M} \tag{2}$$

where C represents the concentration determined from the gallic acid standard curve (mg/mL), V represents the volume of extracts used in the assay (mL) and M represents the mass of extracts used in the assay (g).

2.7. DPPH radical scavenging assay

DPPH is a dark- purple coloured crystalline powder composed of stable free-radical molecules. This assay is used to determine the antioxidant ability of the extract being tested by observing the change in colour from purple to yellow by the reduction process caused by antioxidants present in the extract. The DPPH radical scavenging assay was performed accordingly with minor modifications [5]. An amount of 1.5 mL of 0.1 mM DPPH prepared in methanol was added to 0.5 mL of the sample extracts of different concentrations (0.004 to 0.333 mg/mL) in a test tube. The mixture was vortexed for 30 seconds and left to stand in the dark at room temperature for 30 minutes. Then, the absorbance was measured at 517nm using the Hitachi U-1900 spectrophotometer. The positive control used was ascorbic acid. The negative control was prepared by replacing the extracts with methanol and the blank was prepared by replacing DPPH with methanol. The IC_{50} values were determined using GraphPad Prism and total DPPH radical scavenging ability (%) was calculated using the equation:

$$\text{DPPH radical scavenging ability (\%)} = \frac{Abs(control) - Abs(sample)}{Abs(control)} \text{ x } 100\% \tag{3}$$

2.8. Ferric ion reducing antioxidant power (FRAP) assay

The FRAP assay was performed with slight modifications [13]. In brief, an amount of 3 mL of FRAP working solution was added into 100 µL of plant extracts (0.333 mg/mL) and vortexed. The mixture was placed in a water bath at 37°C for 30 minutes in a dark condition and the absorbance was measured at 593 nm using the Hitachi U-1900 spectrophotometer. Ascorbic acid was used as the positive control. A ferrous sulphate standard curve ranging from 5-100 µM was constructed. The absorbance reading obtained was substituted into the standard curve to obtain the concentration of ferrous sulphate and the results were expressed in FRAP values (µM of Fe (II)/mg of extracts).

2.9. Statistical analysis

All experiments were performed in triplicates and data was presented in terms of mean ± standard error of mean (SEM). The IC_{50} values were calculated using GraphPad Prism. The differences between the data obtained and positive control were assessed using t-test and a p value of less than 0.05 is considered

IOP Conf. Series: Materials Science and Engineering **716** (2020) 012007 doi:10.1088/1757-899X/716/1/012007

as significantly different. Pearson correlation was used to correlate TPC and biological activities of extracts accordingly.

3 Results and discussion

The barks of *Beilschmiedia lumutensis* and *Beilschmiedia insignis* were extracted using two different solvents which were methanol and ethyl acetate. The resulting IC_{50} values of each extracts were calculated and presented in Table 1, where a lower IC_{50} value indicates higher enzyme inhibition activity. The bark methanolic extract of *Beilschmiedia insignis* was observed to exhibit as promising α-amylase and α-glucosidase inhibitor. For α-amylase inhibition, BIM extract had the lowest IC_{50} value of 3.233 ± 0.512 µg/mL, which explains the better potency of BIM extract in inhibiting α-amylase, thus delaying the breakdown and absorption of carbohydrates and lowering the glucose peaks encountered in postprandial glucose [14]. As for α-glucosidase inhibition, BIM extract demonstrated lowest IC_{50} value of 12.357 ± 1.785 µg/mL compared to other extracts.

Plant species are known to contain a broad supply of bioactive compounds or secondary metabolites that are devoted to the beneficial medicinal effects [15]. These bioactive compounds differ in solubility when extracted using different extraction solvents such as methanol and ethyl acetate. Therefore, different active compounds extracted differently by exhibiting different yields when different extraction solvents were used [16]. In the current study, the bioactive compound in the crude extract responsible for α-amylase and α-glucosidase inhibition was more likely to be extracted by methanol. This indicates that the inhibitory activity was influenced by the choice of solvent used during extraction and the solubility of the bioactive compounds.

Table 1. IC_{50} values for α-amylase and α-glucosidase inhibitory activity of two *Beilschmiedia* species (*Beilschmiedia lumutensis* and *Beilschmiedia insignis*) extracted using two different solvents (methanol and ethyl acetate).

Extract	IC_{50} values for α-amylase (µg/mL)	IC_{50} values for α-glucosidase (µg/mL)
BLM	7.398 ± 1.091	58.377 ± 6.675*
BLEA	7.905 ± 1.002	21.913 ± 3.933*
BIM	3.233 ± 0.512	12.357 ± 1.785 *
BIEA	19.203 ± 1.892	46.700 ± 4.963*
Acarbose	13.784 ± 3.703	4148.333 ± 650.177

Each value represent the mean IC_{50} value (N=3, mean ± SEM).
* Values are significantly different ($p < 0.05$) as compared to acarbose.

BIM extract with the lowest IC_{50} value was utilized to further evaluate its mode of inhibition against α-amylase and α-glucosidase activities. The results in the form of Lineweaver-Burk plot were shown in Fig. 1 and 2. It was shown that both the V_{max} and K_m were influenced by the presence of BIM extract which indicated a mixed inhibition. The active compounds that contributed to α-amylase and α-glucosidase inhibition present in BIM extract is believed to bind preferably to the free enzyme instead of the enzyme-substrate complex.

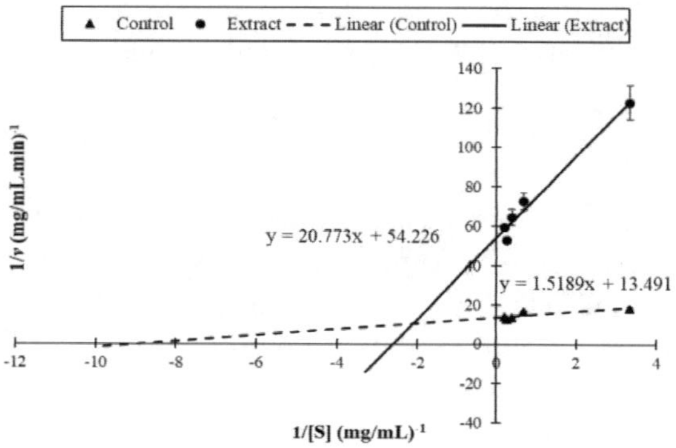

Figure 1. Mode of α-amylase inhibition by BIM extract using Lineweaver-Burk plot.

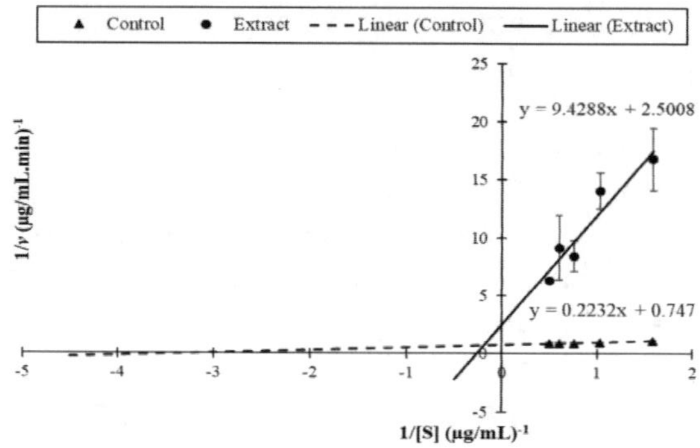

Figure 2. Mode of α-glucosidase inhibition by BIM extract using Lineweaver-Burk plot.

In mixed inhibition, inhibitors tend to attach to the allosteric site of the enzyme instead of the active site. Therefore, it can be concluded that the active compounds in BIM extract interact with the enzyme by binding to the allosteric site of the free enzyme, causing conformational changes to the active site of the enzyme and lowers the enzyme affinity for the substrate, which explains the increase in K_m value. The decrease in V_{max} value observed indicated such inihibition that can be resulted in a retarded breakdown of disaccharides into absorbable monosaccharides, which lowers hyperglycemia peaks in type 2 diabetes [17].

In the present study, the TPC of three *Beilschmiedia* species extracted using two different solvents were determined and shown in Table 2. All four extracts contained polyphenolic compounds. The TPC of all four extracts ranged from 60.439 to 420.393 mg GAE/g extract. The results gathered followed the order: BIM > BLM > BIEA > BLEA. All methanolic extracts had a higher TPC value than their respective ethyl acetate extracts. This result indicated that methanol is a preferred extraction solvent when compared to ethyl acetate in the extraction of polyphenolic compounds due to its higher polarity index. This confirmed that choice of solvent, the polarity index and the solubility of TPC in the solvent affects the recovery of polyphenolic contents [18]. The antioxidant activity of polyphenolic compounds result from their redox properties, enabling them to neutralize free radicals, act as reducing agents, hydrogen donors, quenching singlet and triplet oxygen and metal chelators [12]. Polyphenolic compounds were also reported to contribute towards enzyme inhibition due to their hydroxyl group that is able to bind with proteins, which explains the ability of most *Beilschmiedia* extracts in the current research in inhibiting α-amylase and α-glucosidase [19].

Table 2. Total polyphenolic content of three *Beilschmiedia* species extracted using two extraction solvents, methanol and ethyl acetate.

Extract	Total phenolic content (mg GAE/g extract)
BLM (bark)	344.202 ± 9.275
BLEA (bark)	60.439 ± 4.684
BIM (bark)	420.393 ± 8.010
BIEA (bark)	82.836 ± 7.928

The resulting IC_{50} values for DPPH radical scavenging ability of each extracts were presented in Table 3 in comparison to ascorbic acid (AA). All extracts were weaker than ascorbic acid in scavenging the DPPH radical and had IC_{50} values that were significantly higher ($p < 0.05$) than ascorbic acid. BIM extract with the lowest IC_{50} value of 12.103 ± 0.312 µg/mL demonstrated the strongest DPPH radical scavenging ability whereas BLEA extract with the highest IC_{50} value of 245.867 ± 35.220 µg/mL demonstrated the weakest DPPH radical scavenging ability. The results of the current study proved that TPC of the extracts were obtained in a more polar solvent with increased level of antioxidants. This finding is in line with previous findings where the antioxidant property of the plant extract was correlated to the levels of polyphenolic content and that antioxidant activity of plants were dependent and influenced by the choice of extracting solvents [20]. This finding is of significant as plant extracts with DPPH scavenging ability are capable to donate their hydrogen atom or electrons, acting as primary antioxidants, thus neutralizing the free radicals in hyperglycemia-induced oxidative stress that leads to complications in type 2 diabetes [21]

Table 3. IC_{50} values for DPPH radical scavenging ability of three *Beilschmiedia* species.

Extracts	IC_{50} values for DPPH (µg/mL)
BLM	13.170 ± 0.935*
BLEA	245.867 ± 35.220*
BIM	12.103 ± 0.312*
BIEA	137.950 ± 15.650*
AA	4.930 ± 0.316

* Values are significantly different ($p < 0.05$) as compared to AA.

The reducing potential of all *Beilschmiedia* extracts were evaluated for their ability to reduce the straw colour TPTZ-Fe (III) complex into blue colour, TPTZ-Fe (II). The results revealed that all of the *Beilschmiedia* extracts possessed reducing power, indicated by their respective FRAP values ranging from 499.212 ± 18.884 to 1904.247 ± 24.588 µM Fe (II)/mg extracts (Fig. 3). The results obtained were in the order: AA > BIM > BLM > BIEA > BLEA, which was almost similar to current findings in TPC and DPPH assay. Based on Figure 3, it was observed that both methanolic extracts exhibited higher FRAP values than their respective ethyl acetate extracts, which reconfirmed the trend in our previous findings in TPC and DPPH assay. Highest FRAP value was exhibited by BIM extract. This result was in parallel with a previous study [22], which discovered that the extraction of active compounds was solvent dependent and polar solvent (methanol) is an efficient solvent for extraction of antioxidants and polyphenolic compounds. Therefore, it can be deduced that the high reducing ability of BIM extract was mostly contributed by the polyphenolic compounds present [23].

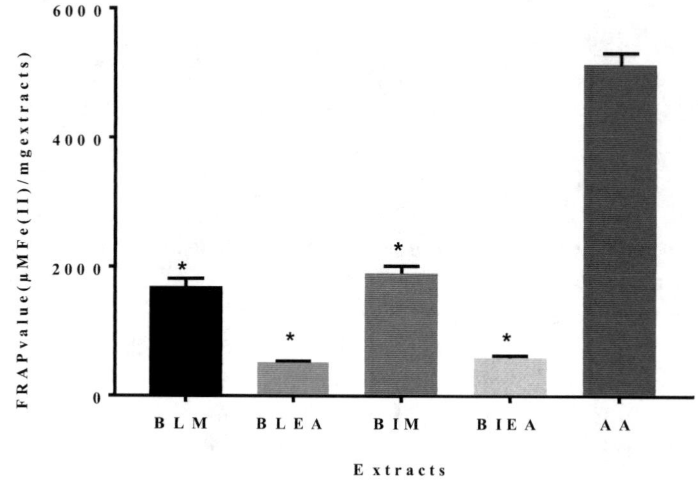

Figure 3. FRAP value of two *Beilschmiedia* species extracted using two extraction solvents, methanol and ethyl acetate. * Values are significantly different ($p < 0.05$) as compared to AA.

Pearson correlation analysis was conducted to analyze the correlations between TPC and some biological activities. Based on Table 4, the highest correlation coefficient, r was found between TPC and FRAP activity and the lowest correlation coefficient was found between TPC and α-glucosidase inhibitory activity. TPC had significant strong correlations with both DPPH and FRAP antioxidant assays with respective correlation coefficient of -0.860 and 0.994. Similar results were reported from a previos report [24], where a strong correlation between TPC and both DPPH and FRAP antioxidant assays were observed with correlation coefficients of 0.921 and 0.868 respectively. This indicated that polyphenolic content found in *Beilschmiedia* extracts used in the current research strongly contributed to the radical scavenging capability and reducing strength seen in both antioxidant assays, which is in line with a previous analysis [25]. TPC and enzyme inhibitory activities were also correlated. it was observed that a higher correlation coefficient (-0.589, $p > 0.01$) was seen between TPC and α-amylase inhibitory activity compared to TPC and α-glucosidase inhibitory activity (-0.191, $p > 0.01$). There is a higher possibility where polyphenolic compounds present in the extracts contributed to α-amylase inhibition than to α-glucosidase inhibition. Therefore, α-glucosidase inhibition activity demonstrated by *Beilschmiedia* extracts in the current study may be contributed by other phytochemicals or secondary metabolites but not in the form of polyphenolic compounds.

Table 4. Pearson correlation coefficient, r between total polyphenolic content (TPC) and the biological activities of *Beilschmiedia* species.

Biological Activities	Pearson correlation coefficient, r
	TPC
DPPH scavenging ability	-0.860**
FRAP ability	0.994**
α-amylase inhibitory activity	-0.589
α-glucosidase inhibitory activity	-0.191

** significant correlation at $p < 0.01$.

Conclusion

It can be concluded that BIM extract serves as a promising candidate for further evaluation as diabetic drug due to strong inhibition against α-amylase and α-glucosidase by mixed inhibition. Along with promising enzyme inhibition potential, BIM extract also possessed antioxidant activity. BIM extract demonstrated the highest TPC content, the lowest IC_{50} in DPPH radical scavenging assay and highest FRAP value out of all tested extracts. This revealed the strong correlation between TPC and both DPPH and FRAP antioxidant assays. Taken together, all these findings supported the potential of methanolic bark extract of *Beilschmiedia insignis* to be part of the potential medicinal plants that may serve as one of the alternative treatments for hyperglycemia-related conditions.

References

[1] Abu Bakar M H, Sarmidi M R, Cheng K K, Ali Khan A, Chua L S, Zaman Huri H and Yaakob H 2015 Metabolomics - The Complementary Field in Systems Biology: A Review on Obesity and Type 2 Diabetes *Mol. Biosyst.* **11** 1742–74

[2] Abu Bakar M H and Sarmidi M R 2017 Association of cultured myotubes and fasting plasma metabolite profiles with mitochondrial dysfunction in type 2 diabetes subjects *Mol. BioSyst.* **13** 1838–53

[3] Hafizi Abu Bakar M, Kian Kai C, Wan Hassan W N, Sarmidi M R, Yaakob H and Zaman Huri H 2015 Mitochondrial dysfunction as a central event for mechanisms underlying insulin resistance: the roles of long chain fatty acids *Diabetes. Metab. Res. Rev.* **31** 453–75

[4] Alam F, Islam M A, Kamal M A and Gan S H 2018 Updates on managing type 2 diabetes mellitus with natural products: towards antidiabetic drug development *Curr. Med. Chem.* **25** 5395–431

[5] Padilla-Camberos E, Lazcano-Díaz E, Flores-Fernandez J M, Owolabi M S, Allen K and Villanueva-Rodríguez S 2014 Evaluation of the inhibition of carbohydrate hydrolyzing enzymes, the antioxidant activity, and the polyphenolic content of Citrus limetta peel extract *Sci. World J.* **2014** 1–4

[6] Nishida S 2008 Taxonomic revision of Beilschmiedia (Lauraceae) in Borneo *Blumea-Biodiversity, Evol. Biogeogr. Plants* **53** 345–83

[7] Salleh W M N H W, Ahmad F, Yen K H and Zulkifli R M 2016 Chemical composition and biological activities of essential oil of Beilschmiedia pulverulenta *Pharm. Biol.* **54** 322–30

[8] Salleh W M N H W, Ahmad F, Khong H Y and Mohamed Zulkifli R 2016 Comparative study of the essential oils of three Beilschmiedia species and their biological activities *Int. J. Food Sci. Technol.* **51** 240–9

[9] Olaokun O O, McGaw L J, Eloff J N and Naidoo V 2013 Evaluation of the inhibition of carbohydrate hydrolysing enzymes, antioxidant activity and polyphenolic content of extracts of ten African Ficus species (Moraceae) used traditionally to treat diabetes *BMC Complement. Altern. Med.* **13** 94

[10] Kazeem M I, Adamson J O and Ogunwande I A 2013 Modes of inhibition of α-amylase and α-glucosidase by aqueous extract of Morinda lucida Benth leaf *Biomed Res. Int.* **2013**

[11] Kim Y-M, Jeong Y-K, Wang M-H, Lee W-Y and Rhee H-I 2005 Inhibitory effect of pine extract on α-glucosidase activity and postprandial hyperglycemia *Nutrition* **21** 756–61

[12] Djeridane A, Yousfi M, Nadjemi B, Boutassouna D, Stocker P and Vidal N 2006 Antioxidant activity of some Algerian medicinal plants extracts containing phenolic compounds *Food Chem.* **97** 654–60

[13] Benzie I F F and Strain J J 1996 The ferric reducing ability of plasma (FRAP) as a measure of "antioxidant power": the FRAP assay *Anal. Biochem.* **239** 70–6

[14] Kotowaroo M I, Mahomoodally M F, Gurib- Fakim A and Subratty A H 2006 Screening of traditional antidiabetic medicinal plants of mauritius for possible α- amylase inhibitory effects in vitro *Phyther. Res. An Int. J. Devoted to Pharmacol. Toxicol. Eval. Nat. Prod. Deriv.* **20** 228–31

[15]B Gaikwad S, Krishna Mohan G and Sandhya Rani M 2014 Phytochemicals for diabetes management *Pharm. Crop.* **5**

[16] Mahomoodally M F and Muthoora D D 2014 Kinetic of inhibition of carbohydrate-hydrolysing enzymes, antioxidant activity and polyphenolic content of Phyllanthus amarus Schum. & Thonn.(Phyllanthaceae) *J. Herb. Med.* **4** 208–23

[17] Kazeem M, Kazeem M I, Oyedapo B F, Raimi O G and Adu O B 2013 Management of Diabetes Mellitus in vitro *J. Med. Sci* **13** 269–75

[18]Abarca-Vargas R, Peña Malacara C and Petricevich V 2016 Characterization of chemical compounds with antioxidant and cytotoxic activities in bougainvillea x buttiana holttum and standl,(Var. rose) extracts *Antioxidants* **5** 45

[19] Mai T T, Thu N N, Tien P G and Van Chuyen N 2007 Alpha-glucosidase inhibitory and antioxidant activities of Vietnamese edible plants and their relationships with polyphenol contents *J. Nutr. Sci. Vitaminol. (Tokyo).* **53** 267–76

[20]Hossain M A and Shah M D 2015 A study on the total phenols content and antioxidant activity of essential oil and different solvent extracts of endemic plant Merremia borneensis *Arab. J. Chem.* **8** 66–71

[21]Matough F A, Budin S B, Hamid Z A, Alwahaibi N and Mohamed J 2012 The role of oxidative stress and antioxidants in diabetic complications *Sultan Qaboos Univ. Med. J.* **12** 5

[22]Barchan A, Bakkali M, Arakrak A, Pagán R and Laglaoui A 2014 The effects of solvents polarity on the phenolic contents and antioxidant activity of three Mentha species extracts *Int J Curr Microbiol App Sci* **3** 399–412

[23]Sadeghi Z, Valizadeh J, Shermeh O A and Akaberi M 2015 Antioxidant activity and total phenolic content of Boerhavia elegans (choisy) grown in Baluchestan, Iran *Avicenna J. phytomedicine* **5** 1

[24] Hyun T K, Kim H-C, Ko Y-J and Kim J-S 2016 Antioxidant, α-glucosidase inhibitory and anti-inflammatory effects of aerial parts extract from Korean crowberry (Empetrum nigrum var. japonicum) *Saudi J. Biol. Sci.* **23** 181–8

[25] Liu H, Qiu N, Ding H and Yao R 2008 Polyphenols contents and antioxidant capacity of 68 Chinese herbals suitable for medical or food uses *Food Res. Int.* **41** 363–70

Acknowledgement

This study was supported by the TWAS-COMSTECH Joint Research Grant-18-420 RG/PHA/AS_C, Universiti Sains Malaysia Short Term Grants (Reference no: 304/PTEKIND/6313329 and 304/PKIMIA/6313330) and RUI USM 1001/PKIMIA/8012310. This research was conducted within the frame of the auspices of CNRS-UM; Associated International Laboratory (LIA) under International French Malaysian Natural Product Laboratory (IFM-NatProLab). We thank Nor, D.M., Syamsir, R., and Teo, L.E. (UM) for the collection and identification of plant material.

IOP Conf. Series: Materials Science and Engineering **716** (2020) 012008 doi:10.1088/1757-899X/716/1/012008

Preliminary study on the growth of *Tetraselmis suecica* in centred-light photobioreactor (CLPBR)

N A Serri[*1], L Anbalagan[1], N Z Norafand[1], M A Kassim[1], M S Abu Mansor[2]

[1]Bioprocess Technology Division, School of Industrial Technology, Universiti Sains Malaysia, 11800, Pulau Pinang, Malaysia
[2]School of Mechanical Engineering, Engineering Campus, Universiti Sains Malaysia, 14300 Nibong Tebal, Pulau Pinang, Malaysia

E-mail: aziah_serri@usm.my

Abstract. Over the past decade, there has been a sustained research on renewable fuel known as biofuel due to the reliance upon fossil fuel. Oil obtained from microalgae is a viable approach for biodiesel production to reduce the reliability on fossil fuel. In this study, the *Tetraselmis suecica* is cultivated in 5L centred-light Photobioreactor (CLPBR) to investigate the effect of pH and light intensity on cell concentration and lipid content of microalgae. Different pH (6.5, 8.5, 10.5) and light intensity (200 lux, 1000 lux, 2000 lux) were used to test their effect on algal growth and lipid production. The results showed that the highest cell concentration (4 mL/mL) and lipid content (18 %) was obtained at pH of 10.5. At 1000 lux light intensity, the highest cell concentration and lipid content was 0.898 mL/mL and 19% respectively. The two models that are tested to determine the growth kinetics are Logistic and Gompertz model. The R^2 obtained from Logistic and Gompertz model was 0.8852 and 0.8736 respectively. The results demonstrate that both models was least accurate to predict the algal growth.

1. Introduction

The use of fossil fuel is unsustainable because of depleting resources, increasing greenhouse gas emission, increased population, and economy crisis. An alternative fuel which is renewable, low greenhouse gas emission, less toxic and low sulphur content preferred [1]. Biodiesel is one of the significant productions of biofuel which is a renewable fuel. The oil produced from microalgae has the ability to substitute the biodiesel. Microalgae are promising feedstock for sustainable production of biofuels and valuable chemicals [2]. The microalgae-based biofuels potentially employed in an economically effective and environmentally sustainable manner. Microalgae is a photosynthetic organism which synthesize biochemical compounds such as proteins, lipids and carbohydrates [3]. Since they are highly resistant organism, they can grow in water sources such as brickish, fresh and salty water [4]. The production of biofuel from microalgae was mainly considered due to its higher biomass and lipid productivity compared to other feedstocks [5].

IOP Conf. Series: Materials Science and Engineering **716** (2020) 012008 doi:10.1088/1757-899X/716/1/012008

In industrial application, the cultivation of algae needs to be in large scale for the high amount of biodiesel production. Thus, few methods were employed for the cultivation of algae such as open ponds, photobioreactor and hybrid system. Photobioreactor plays an important role in the production of environmentally friendly products from microalgae. Photobioreactor was highly recommended for the cultivation of microalgae because it has low risk contamination and the environmental conditions can be controlled [6]. The common types of photobioreactor used for cultivation are tubular, flat panel, airlift and bubble column reactors. Most commonly the lack of innate understanding on light distribution and cell growth in the reactor is one of the limitations in cultivation of microalgae. The most difficulties encountered are the light penetration to reach the cells and overgrowing of algae [7]. Thus, the best conditions for the algae growth and primary factors that leads to negative growth should be analysed carefully to cultivate microalgae.

Kinetic studies are done to predict the growth of algae in a medium. There are different phases in the growth of algae such as lag, exponential, stationary and death phase. The growth kinetic model is robust in describing the growth kinetics [8]. The kinetic growth of algae is usually studied by structured or non-structured models. The models such as Logistic and Gompertz used in this experiment is the example of non-structured models. These models explain as the biomass in the system increases, the specific growth coefficient linearly decreases [9]. The Logistic model is used more frequently than others to study the growth kinetics of algae. The S- shaped curve considered well fitted as the high correlation coefficients are obtained [10]. Gompertz model is also one of the three parameter model which is easier to use for predicting the growth of algae.

There are many researches that has been going on in cultivating microalgae in order to identify best conditions for microalgal growth and lipid production by controlling the light intensity, providing essential nutrients, carbon dioxide and temperature. In this study the growth kinetic of *Tetraselmis suecica* was determined using two kinetic models such as Logistic and Gompertz model. Different parameters such as pH, light intensity and carbon dioxide concentration was investigated to determine the algae growth and lipid production.

2. Materials and method

2.1. Cultivation medium preparation

Cultivation was carried out by using modified algae growth MLA medium that consist of following chemical composition (per litre of distilled water): $MgSO_4.7H_2O$ (49.9 g L^{-1}), $NaNO_3$ (85.0 g L^{-1}), K_2HPO_4 (6.96 g L^{-1}), H_3BO_3 (2.47 g L^{-1}), $NaHCO_3$ (16.9 g L^{-1}), $CaCl_2.2H_2O$ (29.4 g L^{-1}), 1 % (v/v) vitamins solution and 1 % (v/v) micronutrients solution. The vitamins solution was biotin (0.1 g L^{-1}), Vitamin B12 (0.1 g L^{-1}) and Thiamine HCl (0.1 g L^{-1}). The micronutrients solution was Na_2EDTA (4.36 g L^{-1}), $FeCl_3.6H_2O$ (1.58 g L^{-1}), $NaHCO_3$ (0.60 g L^{-1}), $MnCl_2.4H_2O$ (0.36 g L^{-1}), $CuSO_4.5H_2O$ (1.0 g L^{-1}), $ZnSO_4.7H_2O$ (2.2 g L^{-1}) and $CoCl_2.6H_2O$ (1.0 g L^{-1}), $NaMoO_4.2H_2O$ (0.6 g L^{-1})[11]. All chemicals used were analytical grade. Then, the reading of the dried biomass was recorded and cell concentration was calculated by weighing the falcon tubes after drying according to Eq. (1)

$$\text{Cell concentration (g/L)} = \frac{(W1-W2)}{V} \tag{1}$$

Where W_1 is weight of empty falcon tube, W_2 is weight of falcon tube with dried biomass and V is volume of sample

A graph of optical density against cell concentration was plotted. The growth phase of microalgae was determined by recording the absorbance reading at 680nm every 24 hours.

2.2. Growth analysis

The standard curve is done to check the cell concentration by using the absorbance reading. *Tetraselmis suecica* is prepared in six different concentrations with total volume 50 mL in each 50 mL of falcon tube where the absorbance determined at the wavelength 680 nm by (HITACHI U1900) Japan spectrophotometer. The falcon tubes were dried in oven (60 °C) for 24 hours.

2.3. Cultivation of microalgae

The culture was cultivated in MLA medium with controlled environment temperature of 25 ± 3 for 10 days. The cultivation is started by prepared the MLA medium with three different pH (6.5, 8.5 and 10.5) and three different light intensity (200 Lux, 1000 Lux and 2000 Lux) using LED light source. A volume of 1.8 mL of MLA medium and 200 mL of *Tetraselmis suecica* which was cultivated in 1L Erlenmeyer flask added in the CLPBR. Duplicate sample is taken at 24 hours interval for 10 days.

2.4. Biomass harvesting

After 10 days of cultivation, the *Tetraselmis suecica* samples were centrifuged at 5000 rpm for 5 minutes at 4°C. The pellets were rinsed twice with distilled water. The pellets were dried in oven (Binder BD 115) at 70 °C for 24 hours. After that, the dried biomass is crushed into the smaller size and the biomass content is calculated based on microalgae dry weight produced per litre (g/L).

2.5. Lipid extraction

The lipid was extracted from the microalgae biomass by using the Soxhlet extraction. The cell dried biomass was added into the cellulose thimble inside the Soxhlet extraction apparatus unit. A volume of 300 mL n-hexane are used to extract the lipid from the microalgae biomass. The lipid extraction was carried out for 5 hours till the solvent become colourless in the compartment that have cellulose thimble. After the lipid extraction the solvent was evaporated by using the rotary evaporator. The subsequent sample is dried in the oven at 70 °C for overnight. The lipid contents were weighed and determined gravimetrically using Eq. (2):

$$\text{Lipid content (\%)} = \frac{W2 - W1}{m} \text{ X } 100\% \quad (2)$$

Where, W_2 is the weight of the empty beaker and lipid content (g), W_1 is the weight of the empty beaker (g) and m is biomass of the microalgae used for the lipid determination (g).

2.6. Kinetic model

Gompertz model (Eq. 4) and Logistic model (Eq. 5) are used in this project to determine the kinetic behaviour of microalgae growth. Based on these two models, the $R^2 > 0.95$ shows the best fitting to the experimental data for microalgae growth [12]. Specific growth rate, μ was calculated using Eq. (3):

$$\text{Specific growth rate, } \mu = \frac{\ln (X1 / X0)}{t1 - t0} \quad (3)$$

Gompertz model:

$$C(t) = C_o{*}e[-e(\mu{*}e^{1/C_o})]{*}(Cm-t) + 1 \quad (4)$$

Logistic model:

$$C\ (t) = \frac{Co\ e\mu t}{1 - \left(\frac{Co}{Cm}\right)(1 - e\mu t)} \quad (5)$$

Where, , C(t) is the growth kinetics, Co is representing the initial of *Tetraselmi ssuecica* biomass (mg L^{-1}), Cm is maximum biomass, (mg L^{-1}), μ is specific growth rate (day^{-1}) and t is cultivation time (day)

3. Results and discussion

3.1. The growth analysis of microalgae

Figure 1 shows the growth curve of microalgae for 10 days at different pH and light intensity. The growth phase consists of lag, log, stationary and death phase. The lag phase can be called as induction phase where the newly introduced cells in a new culture starts to increase with short lag phase. There is no cell division, however the cells increase in size during this phase. The algal cells are metabolically active at this stage. The algal cells are starting to adapt in a new environment for growth. From the graph it can be observed that starting from day 0 to 2, the algal growth shows lag phase.

From day 3, the algal growth represents the log or exponential phase. At this stage, the cells start to divide, and the algal growth is rapid. The cell density increases as the cells started to adapt the new environment and consume the nutrients in the medium. The cells increase linearly with time as the cell divides at maximum rate [13]. This where the cells are metabolically active. But the duration of exponential phase for different conditions varied. However, usually the growth curve has a stationary phase where the growth curve becomes horizontal because the as the cells divide, the existing cells die at the same time. However, the graph does not project any stationary phase. After the exponential phase it enters the declining phase, where the cell division slows down as the nutrients, pH, carbon dioxide, light becomes limiting factor for growth of algae. This is where the competition for nutrients and other factors increases and the cell become less metabolically active.

In contrast, at pH 7 the cells were still at exponential phase at the end of day 10, which means the cells were still able to grow for longer period due to the nutritional availability in the growth medium. Investigation of *Chlorella* sp. for 12 days showed the growth curve till exponential phase by Wong *et al* [14]. Besides, growth of *Tetraselmis* sp. C7P4 at pH 7 was analysed where at the end of 15th day the microalgae was still at exponential phase [15]. However, another study showed the growth of *Tetraselmis* sp. C7P4 where at the end of 10th day the growth was till stationary phase [16]. The growth curve of light intensity also exhibited the normal growth curve at 200, 1000 and 2000 lux.

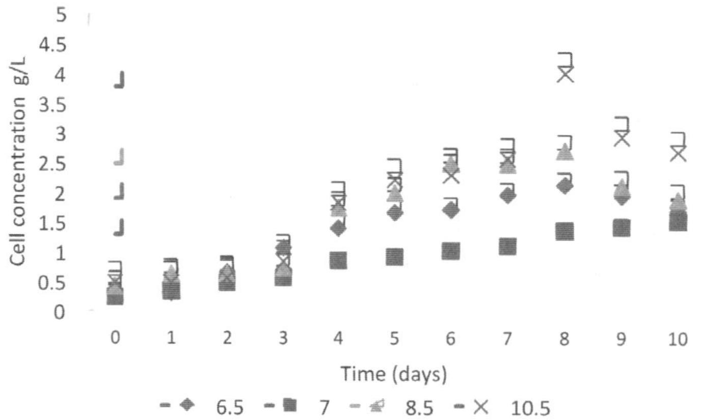

(a)

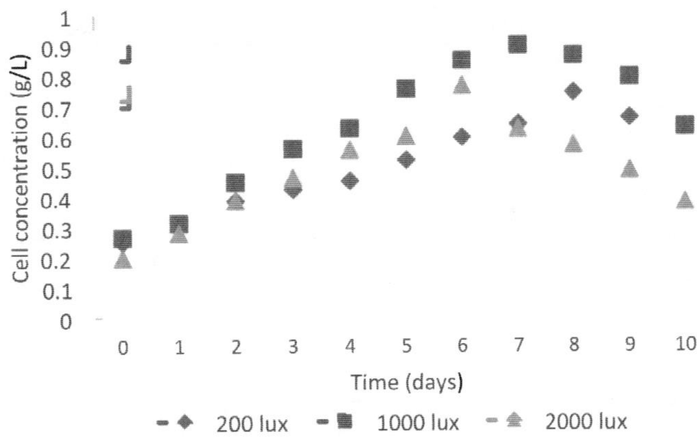

(b)

Figure 1. Cell concentration of *Tetraselmis suecica* at different conditions, a) pH and b) light intensity

3.2. Effect of different conditions on cell concentration and lipid production of microalgae
The variation in pH value in culture can affect cell metabolism and the growth of algae biomass. The pH 7 was used as a control for the cultivation of microalgae. From Figure 1(a), it can be observed that the cell concentration increases as the pH increases. On day 8, the lowest concentration was 1.356mL/mL when the cultivation is done at pH of 7. At pH of 10.5, the highest cell concentration which is 4mL/mL has been recorded. Mostly, the suitable pH for the growth of algae is neutral or alkaline condition. However, the pH required for the cultivation condition depends on the type of strain used. Table 1 shows the highest lipid content was 18% at pH of 10.5and the lowest lipid content was 7.1% at pH of 8.5. It was reported that the alkaline condition can inhibit the microalgae growth and use the energy to form triacylglycerols (TAG) [14]. The cultivation of microalgae in alkaline medium increases the cell wall flexibility. High pH prevents the rupture of cell wall and autospores produced from cell is inhibited. Thus, this will increase the time for the algae to complete growth cell cycle [15].

The cultivation at different light intensity has been conducted in 5L CLPBR 12:12 hour (light : dark) cycle. Figure 1 (b) shows at light intensity of 1000 lux, the highest cell concentration which is 0.898 mL/mL was obtained on day 7. However, using 200 and 2000 lux of light intensity, the cell concentration was the same which is 0.621mL/mL. Light is provided for the microalgae to activate photosynthesis and growth rate. Light distribution to the cells is divided into three distinct zones such as strong illumination zone, weak illumination zone and dark zone [16]. Strong illumination zone causes inhibitory effect for the cells. A study showed that in a flat panel reactor using 1500μmol photon $m^{-2}s^{-1}$, the biomass concentration was 1.8 g/L [17]. In a tubular reactor cultivating *Spirullina* sp. at 2000 lux gave 0.26g dry/L per day [18]. The growth of microalgae depends on the light intensity provided and the type of species used for cultivation. Huang et al.,[19] reported that if the light intensity is very high, the growth of microalgae can be inhibited and cause photoinhibition as the light provided is converted into heat. If the light intensity is low, it can cause photolimitation which eventually kill the microalgae [20]. For lipid content, Table 1 shows the highest lipid content which is 19% was obtained at light intensity of 1000 lux. At 200 lux, the lowest lipid content of 5% is obtained. The difference in light intensity can alter the chemical composition of the microalgae. Low light intensity favours the formation of polar lipids while the high light intensity decreases the polar lipid and increase the neutral storage lipid most probably the triacylglycerol [21]. TAG production under high light conditions might serve as a protective mechanism for the cell [22].

Table 1: The lipid content of *Tetraselmis suecica* under different pH and light intensity

Parameters		Lipid content (%)
pH	6.5	16
	8.5	7.1
	10.5	18
Light intensity (lux)	200	5
	1000	19
	2000	7

3.3. Kinetic model

Growth kinetics *Tetraselmis suecica* was determined by using Logistic and Gompertz models. Gompertz and logistic model was chosen because the growth fitted many organisms such as plant, bird, bacteria and so on. These models are frequently used sigmoid models to fit growth data since it is easy involving less correlated parameters. *Tetraselmis suecica* was grown in MLA medium with the basic environmental conditions such as pH 7, light intensity of 1000 lux, 5% carbon dioxide concentration and temperature of 30˚C. These two models will describe the whole microalgae growth curve. Logistic model represents growth of microalgae based on time, growth rate, initial and final population density [10]. Table 2 represent the experimental growth kinetics where is the R and R^2 for predicted the cell growth kinetics. The R^2 is the coefficient of determination to indicate the model preciseness in fitting the experimental data. The value of R^2 which is close to 1 implies the accuracy of model. From these two models of the Logistic and Gompertz model, the R^2 that was obtained are 0.8852 and 0.8736 respectively. The high correlation coefficients represent the model fitted well to experimental data. However, the correlation coefficient obtained are not so close to 1 indicates that the model is not sufficient to describe the growth of *Tetraselmis suecica*. The low determination of R^2 indicated the Gompertz model is not sufficient to describee biomass production by *Tetraselmis suecica* in medium.

Figure 2 and 3 shows the Logistic and Gompertz model kinetic growth curve. The model might not sufficient enough to fit growth data because of its many re-parameterisations applied in different field of study for different types of growth [23]. These models do not suit to study growth data of *Tetraselmis suecica*. Similar result was found in the investigation of *Chlorella vulgaris* where the Gompertz and Logistic model does not describe the accuracy of data [24]. Lim et al.,[24] studied the growth kinetics of *Chlorella* sp. using Richards, Gompertz and Logistic model. Richards model represents a good fit of data compared to Logistic and Gompertz model. However, a study conducted by Abinandon et al., described the growth kinetics efficiently using both Logistic and Gompertz model with *Chlorella* sp [25].

Table 2: The kinetic growth model of microalgae

Parameters	Models	
	Logistic	Gompertz
R	0.9409	0.9346
R^2	0.8852	0.8736

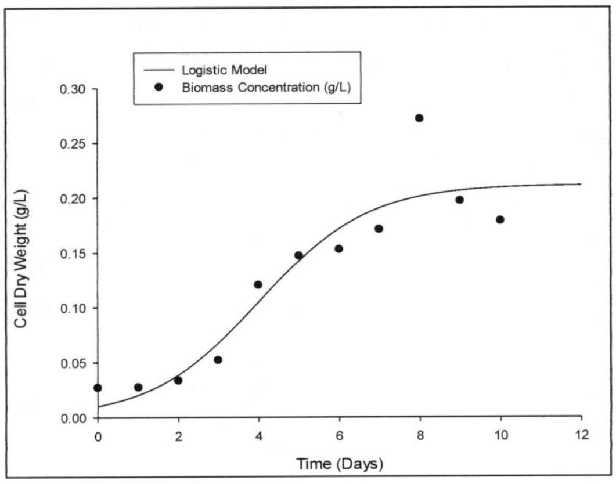

Figure 2: Logistic model for *Tetraselmis suecica* kinetic growth.

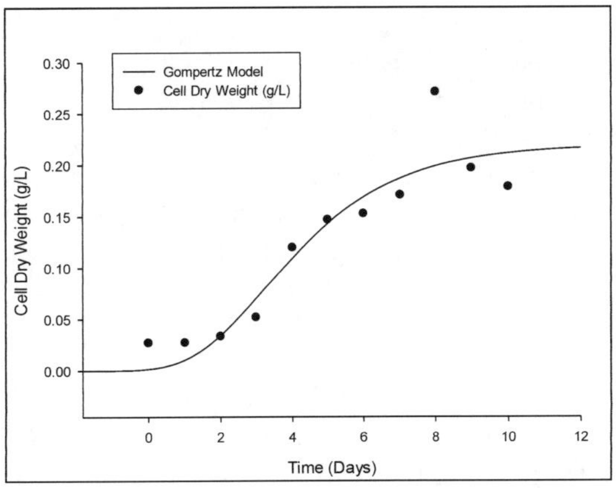

Figure 3: Gompartz model for *Tetraselmis suecica* kinetic growth.

4.Conclusion

In conclusion, this study showed that the growth of algae and lipid production is greatly influenced by the difference in pH and light intensity. The results revealed that the growth profiles are different for *Tetraselmis suecica* under these two conditions. From this study, the best pH for cultivation of *Tetraselmis suecica* in 5L CLPBR is 10.5 which gives the highest cell concentration and lipid content. Besides, the highest cell concentration and lipid content observed at 1000 lux of light intensity. It indicates that the light intensity of 1000 lux is more suitable for growth and lipid production of *Tetraselmis suecica*. The two growth kinetic models such as Logistic and Gompertz that have been tested gives R^2 that is not so close to 1. It shows that both Logistic and Gompertz model is least to predict the growth kinetics of *Tetraselmis suecica* in 5 L CLPBR. For future recommendation to predict the growth of microalgae, the Logistic and Gompertz model that is modified can be used instead for better performance.

5. References

[1] Hartmann D L Klein T A M G Rusticucci M 2013 Fifth Assessment Report - Climate change in J Hurrell ed. IPCC.
[2] Singh R and Sharma S 2012 Development of suitable photobioreactor for algae production – A review *Renewable and Sustainable Energy Reviews.* 16(4) 2347- 53
[3] Richmond A 2004 Handbook of microalgal culture: biotechnology and applied phycology Blackwell science Ltd. 198- 211
[4] Acie´n F G, Ferna´ndez J M, Maga´n J J, Molina E 2012 Production cost of a real microalgae production plant and strategies to reduce it *Biotechnol Adv.* 30 1344–53
[5] Mehta A & Mehta N 2015 Algal biofuel: futuristic trends in fuel industry *International research journal of engineering and technology*. 2(5) 69-70
[6] Zittelli G C, Blondi, Rodolfi L, Tredici M R 2013 Handbook of Microalgal Culture: Applied Phycology and Biotechnology *John Wiley and Sons*. 225-51
[7] Bilos L Patyna A Placzek M Witczak S 2016 Cultivation Of microalgae *chlorella vulgaris* in laboratory photobiorecator *Economic And Environmental Studies*. 16(4) 843-52
[8] Mohamed M S Tan J S Kadkhodaei S Mohamad R Mokhtar M N Ariff A B 2014 Kinetics and modeling of microalga *Tetraselmis sp. FTC 209* growth with respect to its adaptation toward different trophic conditions *Biochemical Engineering Journal.* 88 30–41

[9] Biria D E Maghsoudi & Roostaazad R 2014 Application of power Law logistic model to growth kinetics of *Bacillus licheniformis ms3* on a water in soluble substrate, *Chemical Engineering Communications*. 201(11) 1514-25

[10] Tjørve KMC, Tjørve E (2017) The use of Gompertz models in growth analyses, and new Gompertz-model approach: An addition to the Unified-Richards family *PLoS ONE*. 12(6) 2 13

[11] Esson D Wood S A & Packer M A 2011 Harnessing the self-harvesting capability of benthic cyanobacteria for use in benthic photobioreactors. *AMB Express*. *1*(1) 19.

[12] Celekli A Yavuzatmaca M Bozkurt H 2009 Modeling of biomass production by *Spirulina platensis* as function of phosphate concentrations and pH regimes *Bioresource Technology*. 100 3625–29

[13] Farag I Price K 2013 Resources Conservation in Microalgae Biodiesel Production *International Journal of Engineering and Technical Research (IJETR)*. 1 49-56

[14] Wong Y K Ho Y H Ho K C Leung H M Yung K K L 2017 Growth Medium Screening for Chlorella vulgaris Growth and Lipid Production *J Aquac Mar Biol*. 6(1) 2-10

[15] Pereira H Paramo J Silva J Marques A Barros A Mauricio D Varela J 2018 Scale-up and large-scale production of *Tetraselmis* sp. CTP4 (Chlorophyta) for CO2 mitigation: from an agar plate to 100-m3 industrial photobioreactors *Scientific Reports*. *8*(1) 2-9

[16] Pereira H Katkam G Schulze P S Tamara S Carolina B S Lisa M S Custodio L Malcata X F Gouveia L Varela J C S Barreira L 2016 Isolation of a euryhaline microalgal strain

[17] Li J Stamato M Velliou E Jeffrys C Agathos S N 2014 Design and characterization of a scalable airlift flat panel photobioreactor for microalgae cultivation *J Appl Phycol*. 2-12

[18] Mahmoud R Ibrahim M Ali G 2016 Closed Photobioreactor For Microalgae Biomass Production Under Indoor Growth Conditions *J Algal Biomass utln*. 7(1) 86-92

[19] Huang Q, Jiang F, Wang L, Yang C 2017 Design of Photobioreactors for mass cultivation of photosynthetic organisms *Engineering*. 3(3) 318-29

[20] Zhuang L, Yu D, Zhang J, Liu F, Wu Y, Zhang T, Hu H 2018 The characteristics and influencing factors of the attached microalgae cultivation: a review *Renewable and Sustainable Energy Reviews*. 94 1110-19

[21] Hassan F M Aljobory I F Al-Jumaily E F 2013 Stimulation of biodiesel production from two algae:*Chlorella vulgaris* Berjerinck and Nitzschiapalea (Kütz.) Smith and study their some growth parameters under different light intensity IOSR JESTFT. 6(2) 31-42

[22] Lv J M Cheng L H Xu X H Zhang L Chen H L 2010. Enhanced lipid production *of Chlorella vulgaris* by adjustment of cultivation conditions *Bioresource Technology*. 101 6797–6804

[23] Mansouri. M 2016 Predictive modeling of biomass production by *Chlorella vulgaris* in a draft tube airlift photobioreactor Advances *in Environmental Technology*. 3 119-26

[24] Lim B R Jutidamrongphan K Y 2010 Comparison of Models to Describe Growth of Green Algae *Chlorella vulgaris* for Nutrient Removal from Piggery Wastewater *AGRIS*. 6(52) 19-26

[25] Abinandan Praveen Natarajan Kavitha M. S Biochemical responses from biomass of isolated *Chlorella* sp under different cultivation modes: non-linear modelling of growth kinetics 2018 *Brazilian Journal of Chemical Engineering*. 35(2) 489-96

Acknowledgement

This research is supported by Universiti Sains Malaysia under Bridging grant (Reference No.: 6316214).

IOP Conf. Series: Materials Science and Engineering **716** (2020) 012009 doi:10.1088/1757-899X/716/1/012009

Design and development of centred-light photobioreactor for microalgae cultivation system

N S Mustapa[1], M S Abu Mansor[1,*] and N A Serri[2]

[1] School of Mechanical Engineering, Engineering Campus, Universiti Sains Malaysia, 14300 Nibong Tebal, Pulau Pinang, Malaysia.
[2] School of Industrial Technology, Universiti Sains Malaysia, 18000, Pulau Pinang, Malaysia.

*Email: mesalman@usm.my

Abstract. The cultivation of microalgae within a controlled condition allows the production of biomass with lower contamination and suitable for high value pharmaceutical products. The controlled environment for microalgae growing can be accomplished by utilizing photobioreactor as a platform for introducing the basic needs needed by microalgae to grow effectively. However, due to some problems of high cost, low illumination and complicated cultivation systems which referring to available photobioreactors had contributed to low biomass production. This study deals with designing a new lab scale photobioreactor for overcoming the problems contributed by the available photobioreactors. During the developing process of the photobioreactor, the user needs are identified, and the most suitable concept based on the user needs was generated. The concept was then fabricated to prototype and its performance was validated. A centred-light photobioreactor was successfully developed for lab scale usage and being tested by cultivating *Chlorella vulgaris* species. A maximum of 0.281 g/L cell concentration was achieved using white light source during cultivation of *Chlorella vulgaris*.

1. Introduction

Microalgae can be grown in the two types of systems which are either open systems, such as raceway ponds or closed systems, such as photobioreactors [1]. A closed equipment which utilizing closed system is called as photobioreactor is useful for providing a controlled environment which able for enabling the high productivity of microalgae [2]. There are various commercial applications for microalgae nowadays which are enhancing the nutritional value of animal feed and human food, helping in aquaculture and improving cosmetics production [3]. In an open system, the growing of microalgae is highly potential to be exposed to contamination. These conditions occurred because the environment for microalgae growing cannot be controlled as in closed systems.

Recently, various types of photobioreactors have been designed and developed for production of algae. Photobioreactors can be placed indoor or outdoor. Indoor photobioreactor usually having lab scale size which is smaller than outdoor photobioreactor and suitable to be used by researches.

Outdoor photobioreactors are suitable for large scale production of microalgae and mostly used by industries. There are few standard designs of photobioreactors that can be used as a foundation for further improvisations which are categorized as flat plate reactors, annular reactors and tubular reactors [2,4]. Annular reactor usually acts as bubble column which the inner cylinder is empty to increase the surface to volume (S/V) ratio and to avoid any dark parts [4]. Among these reactors, tubular reactors which having transparent tubes are more preferable and commonly used in photobioreactors design. There are big numbers of closed photobioreactors designed are made by using tubular reactor [4]. In designing the tubular reactors, the choice of the tube diameter is becoming an important decision for getting the optimal design of the photobioreactors. The choice of tube diameter affects the surface to S/V ratio which resulting the light uptake of the culture [5].

There are quite a number of lab scale photobioreactors that have been in the market but most of the available photobioreactors are having a high cost, low illumination surface area and complicated algae harvesting systems. Photobioreactor systems allow for better control of the algae culture environment but the systems tend to be more expensive than the raceway ponds [6]. Chun-Yen et al. [7], argue that although many efforts have been made to develop an efficient and cost-effective of photobioreactor, however, the high cost of installing and operating artificial light sources in conventional photobioreactors with artificial illumination systems remains a major problem. Mass production of microalgae oil faces a number of technical problems that render the current development of the algae industry economically unfit [7]. In addition, it is also necessary but very difficult, to develop cost-effective technologies that would permit efficient biomass harvesting and oil extraction [7]. Thus, these limitations in available photobioreactors must be eliminated to exploit the potential in algae industries.

The study consists of designing a lab scale photobioreactor that is simple and reliable to overcome the limitations of high cost, low illumination surface area and complicated algae drainage system in available photobioreactors for microalgae growing. The photobioreactor should have minimum parts as to reduce the cost. Moreover, the photobioreactor should allow high penetration of light to cultivation medium as to increase the illumination. In order to allow the high productivity of algae, sufficient amount of CO_2 and light supplied should also be considered in designing the new photobioreactor.

2. Methodology

2.1 Design of cultivation system

A lab scale of photobioreactor which suitable to be used for microalgae is developed based on demands from a user in an Alpha Institution. The demands from customer then being translated to the desired specifications for the lab scale photobioreactor. From the specifications, several concepts are generated and the best concept was chosen to develop as a final product. Computer Aided Design, CAD model or 3D model for the chosen concept was then created by using SolidWorks software. The model was produced in order to give more understanding about concept. The chosen concept which has problems was modified for improvements and the CAD model for the concept which has been improved was then created. Moreover, the prototype of developed photobioreactor consisted of several parts, some parts are fabricated with some parts are standard. Once the prototype is completed, the performance for the photobioreactor was studied. The flowchart for this study is shown in Figure 1.

2.2 Fabrication of cultivation system

The materials needed for fabricating parts in cultivation system are prepared as in Table 1. The process for fabrication is begun by making holes. Holes as markers with diameter of 5 mm are made on the lids and tanks of cultivation followed by the holes with respective required diameters (12.5 mm, 13 mm, 16.5 mm, 15 mm and 21 mm) by using bench drilling machine with rotation of 340 rev/min. The lids, transparent pipes, ball valves, tank nipples, clippers, rubber washers, nozzles and tank connectors are then assembled to the cultivation tanks. There are 3 units of cultivation tanks and each tank contains 1 unit lid, 2 units (260 mm length, 22 mm outer diameter and 19 mm inner diameter) transparent pipes, 1 unit ball valve, 1 unit tank nipple, 2 units clippers, 6 units rubber washers, 1 unit nozzle and 2 units tank connectors. Sealing tape is used if there is any leakage to the tanks.

For the upper base, the fabrication processes are begun by cutting the 15 mm width × 15 mm height square mild steel tube to the lengths of 180 mm (2 units), 405 mm (2 units) and 150 mm (4 units) by using hacksaw. Angles of 45° are made at the one ends of the 150 mm length (4 units) and both ends of 80 mm length (2 units) of the square mild steel tube by using hacksaw. A mild steel plate with the dimension of 37 mm width × 3 mm thickness is cut by using shear cutter machine to 575 mm length (3 units). Then, the mild steel plates are bent to the shape of oblong. Another 2 units of mild steel plate with the dimension of 175 mm length × 17 mm width × 1.2 mm thickness are prepared by using shear cutter machine and 2 holes with diameter of 6 mm are made on the plates by using bench drilling machine. The oblong shapes, plates and the prepared square mild steel tubes are assembled together by using metal inert gases, MIG welding to form a part called upper base. In order to remove the excessive weld material formed on the parts, hand grinder is used.

Table 1 Materials needed for fabricating photobioreactor system.

Parts	Materials and descriptions
Lids (3 units)	• Made from polypropylene plastic • Diameter for holes: - 12.5 mm (1 hole) - 21 mm (2 holes) - 16.5 mm (1 hole)
Cultivation tanks (3 units)	• Made from polypropylene plastic • Diameter for holes: - 15 mm (2 holes) - 13 mm (1 hole) • Standard parts used: - 22 mm outer diameter and 19 mm inner diameter of transparent pipes (6 units of 260 mm length) - Ball valves (3 units) - Tank nipples (3 units) - Clippers (6 units) - Rubber washes (9 units) - Nozzles (3 units) - Tank connectors (6 units) - Sealing tapes
Upper base (1 unit)	• 15 mm width × 15 mm height of square mild steel tube - 180 mm length (2 units) - 405 mm length (2 units) - 150 mm length (4 units) • 37 mm width × 3 mm thickness of mild steel plate - 575 mm length (3 units) • 17 mm width × 1.2 mm thickness - 175 mm length (2 units)

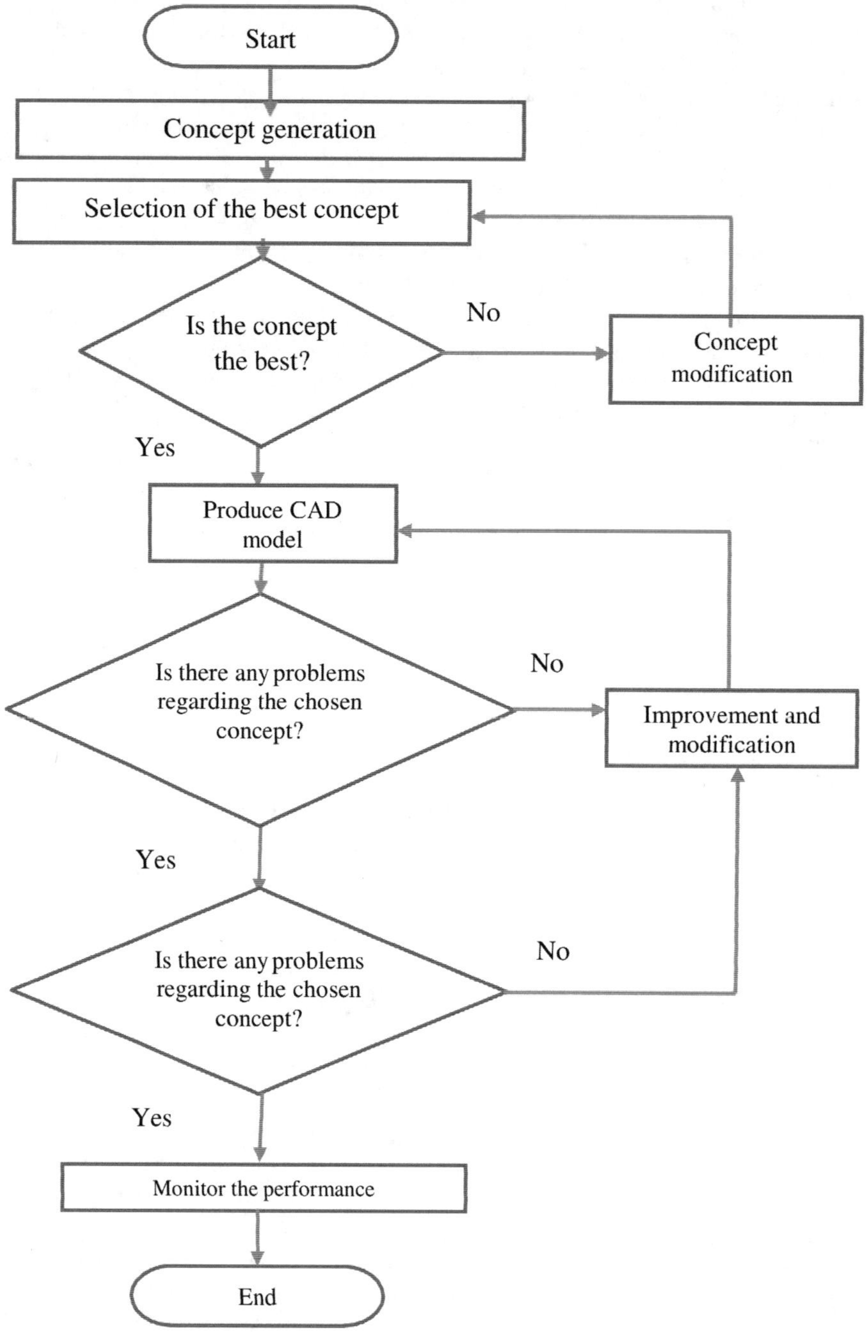

Figure 1 Flowchart for developing a new lab scale of photobioreactor.

IOP Conf. Series: Materials Science and Engineering **716** (2020) 012009 doi:10.1088/1757-899X/716/1/012009

2.3 Performance monitoring stage or testing

The performance of developed prototype for the centred-light photobioreactor was investigated using *Chlorella vulgaris*. The cultivation of *Chlorella vulgaris* was done using MLA medium (modified algae) growth medium [8] at pH 10.5. The vessel filled with 3.6 litre of medium mixed 400ml of microalgae (starting culture) was added to start the process. Two type of light source was used; white color and red color light were set to 3000 lux. Triplicate sample were collected at 24hour interval for 12 days. The biomass concentration was then determined by measuring optical density at 680nm using spectrophotometer.

3. Results and discussion

3.1 Prototype of lab scale photobioreactor

Table 2 shows the final concept which has been chosen to develop the prototype. It provides more advantages if compared to conventional method using shake flask or 5.0 litre mineral bottle setup. A 13500 ml photobioreactor (each tank has volume of 4500 ml) consists of cultivation system (Table 2) is developed and fabricated. The overall dimension of the photobioreactor is 440 mm length × 230 mm width × 740 mm height. The parts in cultivation system are attached by screws to parts in drainage port and they can be removed for convenience. Light-emitting diodes (LED) lights are standard parts which have red colour of lights. The air pump is also one of the standard parts used which has flow rate of 3000 ml/min. Only one air pump is used in the photobioreactor and all the tanks need to share. Thus, each tank will have 1000 ml/min. The illumination system that used centre-lighting was integrated within the system to have a wider range of light emission in the reactor (0- 2000 lux) and can be easily change with different type of colours (red and white). Vo et al (2019) stated illumination patterns and different source of light can be attractive solutions to increase biomass yield in photobioreactor design [9].

Table 2: Cultivation system.

Cultivation system	Function
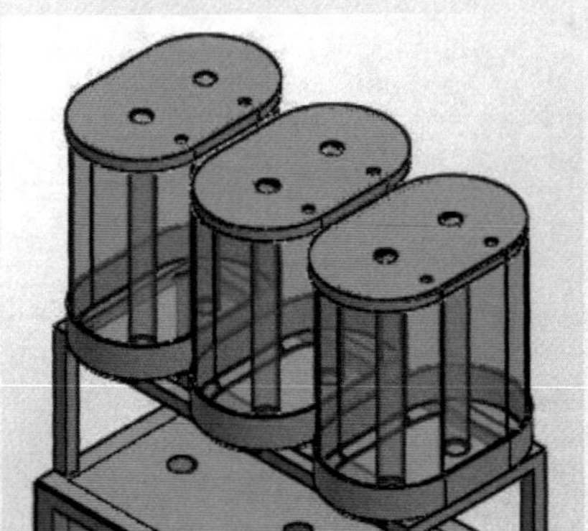 The final concept that has been chosen to develop to prototype.	The function for cultivation system is to be a vessel for growing microalgae. The basic needs required by microalgae are introduced in this system. **Problems solved:** • Advantages 1: One of the main advantages is to prevent from water blockage at the cultivation tanks since ball valves and nozzles are used to transfer water and microalgae from cultivation tanks via drainage port. • Advantages 2: Another main advantage is that the working volume is higher compared to previous concepts and thus allows for high productivity of algae. • Advantages 3: The tanks are separated to each tank and has a working volume of 4500 ml. It is because to avoid contamination. If one of the tanks has been contaminated, it will not affect another tank. Thus, operating cost can be reduced.

3.2 Cultivation of Chlorella vulgaris in prototype photobioreactor

The effect of white light and red light were tested on the microalgae growth. Figure 2 shows the effect of microalgae growth using different light colors on *Chlorella vulgaris*. The white light gave higher cell concentration, 0.281 g/L compared to red light cultivation, 0.197 g/L which is 1.42 times higher. However, according to previous study by Rai and co-workers (2015) red light give higher cell concentration if compared to white light [10]. The result might vary due to different type of light source (LEDs) and positioning of the light source. Too much light can damage the microalgae or inhibit the growth.

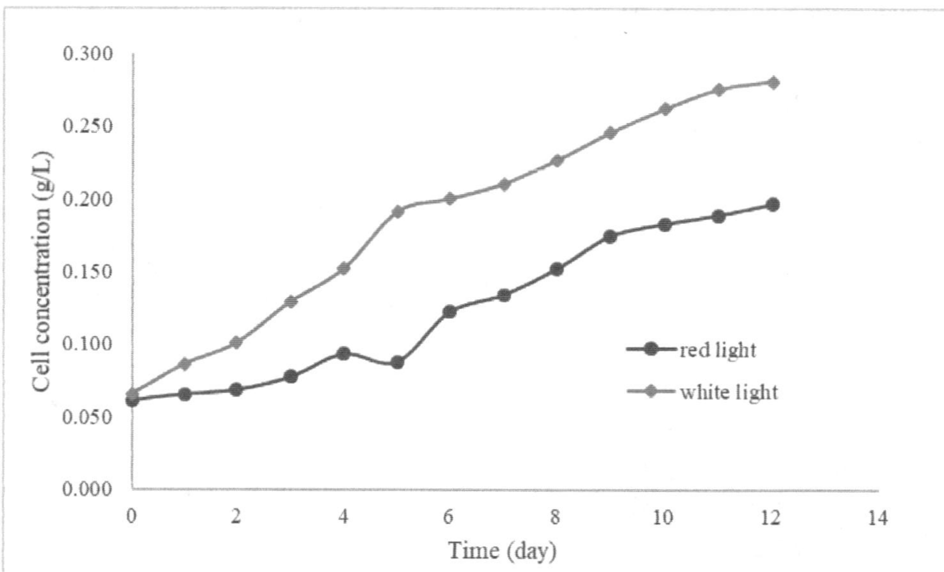

Figure 2: *Chlorella vulgaris* growth in prototype photobioreactor using 2 different light sources.

4. Conclusion

The development of new design of photobioreactor has been successfully executed especially for the cultivation system. Even the designed photobioreactor is not fully overcome the problems faced by the end user such as in-situ sterilization and fully automated controller, but it still provides the user with easy drainage port, better light illumination and bigger operating volume for higher productivity with minimum space required. The photobioreactor also shows a growth of *Chlorella vulgaris* cultivation in 12 days.

IOP Conf. Series: Materials Science and Engineering **716** (2020) 012009 doi:10.1088/1757-899X/716/1/012009

5. References

[1] Sforza E, Bertucco A, Morosinotto T and Giacometti G M 2012 Photobioreactors for microalgal growth and oil production with Nannochloropsis salina: From lab-scale experiments to large-scale design. *Chem Engine Res and Design*, **90(9)** pp 1151-1158

[2] Siddiqui S, Rameshaiah G N and Kavya G. 2015 Development of Photobioreactors for Improvement of Algal Biomass Production. *I J of Scientific Res*, issue 1 Jan 2015, **4** pp 220-226

[3] Spolaore P, Joannis-Cassan C, Duran E and Isambert A 2006 Commercial applications of microalgae. *J Biosci Bioeng,* **101(2)** pp 87-96.

[4] Schenk P M, Thomas-Hall S R, Stephens E, Marx U C, Mussgnug H J, Posten C, Kruse O and Hankamer B 2008 Second Generation Biofuels: High-Efficiency Microalgaefor Biodiesel Production. *Bioenergy Res,* **1** pp 20-43

[5] Torzillo G and Chini Zittelli G 2015 Tubular Photobioreactors. *Algal Biorefineries* ed A Bajpai R and Zappi M (Springer, Cham) pp 187-212

[6] Slade R and Bauen A 2013 Micro-algae cultivation for biofuels: Cost, energy balance, environmental impacts and future prospects. *Biomass and Bioenergy,* **53** pp 29-38

[7] Chen C Y, Yeh K L, Aisyah R, Lee D J and Chang J S 2011 Cultivation, photobioreactor design and harvesting of microalgae for biodiesel production: a critical review. *Bioresour Technol,* **102 (1)** pp 71-81

[8] Bolch C J S and Blackburn S I 1996 Isolation and purification of Australian isolates of the toxic cyanobacteruium *Microcystis aeruginosa* Kütz *J Appl Phycol* **8 (1)** pp 5-13

[9] Vo H N P, Ngo H H, Guo W, Nguyen T M H, Liu Y, Liu Y, Nguyen D D and Chang S W. 2019 A critical review on designs and applications of microalgae-based photobioreactors for pollutants treatment. *Science of the Total Env*. Vol 651 pp 1549-1568.

[10] Rai M P, Gautom T and Sharma N 2015 Effect of salinity, pH, light intensity on growth and lipid production of microalgae for bioenergy application. *OnLine J of Biol Sciences*, **15 (4)** pp 260-267

6. Acknowledgements

This research is supported by the Ministry of Higher Education Malaysia and Universiti Sains Malaysia under the Fundamental Research Grant Scheme (FRGS) (Reference No.: 6071369) and Research University Grant (Reference No.: 814247).

IOP Conf. Series: Materials Science and Engineering **716** (2020) 012010 doi:10.1088/1757-899X/716/1/012010

A study on design improvement for tank photobioreactor

M W Lee[1], M S Abu Mansor[1,*] and N A Serri[2]

[1] School of Mechanical Engineering, Engineering Campus, Universiti Sains Malaysia, 14300 Nibong Tebal, Pulau Pinang, Malaysia.
[2] School of Industrial Technology, Universiti Sains Malaysia, 18000, Pulau Pinang, Malaysia.

*Email: mesalman@usm.my

Abstract. This study aims to design an improved photobioreactor from the current conventional approached in cultivating microalgae in lab scale using conical flask and glass tanks. In order to design a photobioreactor, thus the study required to investigate parameters that important affecting the cultivation process of the microalgae. The study also applied several techniques and knowledge of product design during the designing process for the photobioreactor. It is also important to consider the present product to be used for the industrial scale in future, therefore the techniques for product design must be applied. The techniques used are such as data collection of customer needs, concept generation, concept screening, concept selection and others knowledge also have been used. Finally, the developed photobioreactor was validated by cultivating 4 litre of *Chlorella sp.* within 10 days in MLA medium at pH 10. The highest cell concentration achieved at 0.33 g/l after 4 days of cultivation.

1. Introduction

Microalgae are the longest living autotrophic taxon in this world, and they are tissue-less organisms, simple multicellular organisms that can reproduce sexually or asexually [1]. Microalgae can be used to produce the biofuels because they synthesize three main biochemical compounds which are carbohydrates, proteins and lipids in large amount [1]. Those biochemical compounds are the main compounds to produce biofuels. Algae is more suitable than other plants in cultivating to produce biofuels because they produce large amount of those three compounds, and it is possible for them to produce 10 to 100 times of more oil compared to those oil plants [1]. Since algae are so important for the production of biofuels, it should be cultivated and can be handled easily to fulfil the demand for the renewable energy because of rapid population growth in future. Algae can be cultivated in two types of systems which is open system such as pond or lake and closed system which is known as photobioreactor [1-4].

Photobioreactor can be described as an enclosed, illuminated culture vessel designed for controlled biomass production [1]. The photobioreactor are usually made of glass, poly vinyl chloride (PVC), low-density polyethylene (LDPE) or other materials that are transparent [2]. Photobioreactor are usually used for the cultivation of microalgae because it is a closed system, so the environment or condition inside the photobioreactor can be easily controlled by the scientists. Besides, photobioreactor also prevent the system from contaminants which will destroy the growth of microalgae in the system. Other than that, the photobioreactor also provides the advantages over the open system where photobioreactor can achieve higher productivity per unit area, allow to save water and chemical agents, and photobioreactor is suitable for outdoor or indoor installation [1]. However, there are many parameters that will affect the performance of the photobioreactor such as light, mixing, temperature, and mass transfer. Besides, the photobioreactor that available now is not preferable for the mass production of algae due to many factors such as cost, design and especially the productivity that affected by the design of the photobioreactor and also due to lack of deep understanding in the parameters mentioned as well as deeper understanding of efficient reactor design [5]. Therefore, this study is to design the tank photobioreactor that can help to improve the productivity of biomass production.

Microalgae are photoautotrophs and can use light as their only energy source for the synthesis of biomass and metabolites in photoautotrophic conditions. Therefore, light is one of the most critical parameters for the cultivation of microalgae, thus the design of photobioreactor needs to consider the light intensity for the optimal condition because the growth rate for different species of microalgae will affected by different light intensity. There is a research have been done shows that microalgae were able to perform photosynthesis at a higher efficiency under the low light conditions. The reason microalgae were not able to achieve high photosynthetic efficiency under high light intensity because microalgae need to dissipate part of the absorbed light energy as heat instead of biochemical energy [6]. Although the research shows that low light intensity is better for the cultivation of microalgae, but the optimal light intensity for microalgae is varies significantly among different algal species.

Besides, the light emitted will also emit the heat to the photobioreactor at the same time. Some researchers have done the research on the relationships between light exposure and broth temperature. The study shows that the reactor wall and shading factor where both related to direct and reflected solar irradiation are the major influence on the broth temperature. From the research, the researchers conclude that the best way to maintain temperature range is limiting the irradiance reaching the reactor [7].

On the other hand, the light exposure not only affects the temperature for microalgae, it also involves the photoperiods of microalgae. Photoperiods are the periods of the microalgae exposed to the light and it can be classified into three main groups which are long-term, frequency and short photoperiods. Long-term photoperiods encompass periods of hours, frequency photoperiods are number of cycles per day while short photoperiods are cover light/dark cycles in seconds or milliseconds. The results of the research show that *Scenedesmus obliquus* species of microalgae was able to store enough energy for the cell growth for continuous periods up to 2 hours in the dark without affecting the photosynthetic rate. Therefore, the application of knowledge of photoperiods in cultivation of microalgae can reduce the energy demand by microalgae during photosynthetic process. Thus, it is a very effective strategy in reducing costs for biomass production in condition the problem of only natural photoperiods available in real cultivation systems are overcame [8].

Furthermore, technical improvements in algae-based biofuel production have been magnified in recent years [9]. The artificial cultivation system known as photobioreactor also is the most effective technology for cultivation of microalgae, in terms of high production volume because of the improved cultured environmental conditions. However, the economic feasibility is still questionable due to the high operating cost, which makes generic engineering cultivation method a potential innovation, but the environment feasibility of this method is under heavy criticism [10]. Thus, this study aims to develop a prototype for a lab scale algae cultivation tank with better control of light emission and air flow. *Chlorella sp.* species was used to validate the ability of the prototype developed.

2. Methodology

2.1 Design and Simulation

Observation is required so that further understanding is achieved instead of by using imagination on the existing products for identifying customer needs. The customer in this study is defined as the end user of photobioreactor. Besides, the observation of existing products also exposed the working principles of photobioreactors which is important in the design stage. Besides, an interview with customer is carried out to understand and investigate the customer needs for this study. The questionnaire from interview with customer is developed based on the existing products.

The list of needs according to customer has been created for generating product specifications. Second, list of metrics is also been created to list out all the specifications that required for the design of a photobioreactor. Then, a needs-metrics matrix was created according to the list of metrics to determine the functionality of each specifications and the necessity of considering that specifications during the design of the photobioreactor.

In concept generation stage, the concept of product is generated. During the generation of concept for product, the details of concept have been searched externally to have an exposure of the design available through literature review. Then, the concept also has been searched internally which means this stage will use the personal knowledge and creativity to generate a lot of solution concept. Screening process is performed in concept selection stage by using the concept screening matrix to screen the concept into two most promising concepts from the large amount concept generated in previous stage. Then, the last two concepts undergone the concept scoring matrix in order to select the most promising concept and the concept will be selected for further development.

For conceptual design stage, the final conceptual design of the laboratory scale for the concept of photobioreactor has been created by using the SolidWorks software. SolidWorks software also has been used to do the analysis for the assembly, interferences, clearance, 2D drawing, and exploded view of the model. Besides, the performance of the design such as simulation also been studied and performed using the SolidWorks software.

2.2 Fabrication

After the simulation is performed by using SolidWorks software, the fabrication process is carried out according to the conceptual design in software. Several tools have been used in this fabrication process such as the plastic cutter to cut the Perspex used for the design, chloroform liquid which used to glue the Perspex together and others. However, since some of the parts and components are available in them market, therefore procurement process is taken instead of fabrication process. The product which is photobioreactor is fabricated and assembled according to the conceptual design. However, only the main body of the photobioreactor is fabricated meanwhile other components or parts are bought and obtained from market and recycled from industrial wastes.

2.3 Validating stage with Chlorella sp.

The developed cultivation vessel was tested using *Chlorella sp*. The cultivation was done using MLA medium (modified algae) growth medium at pH 10. The whole photobioreactor was sterilised under UV light inside the laminar flow for 20 minutes. The vessel filled with 90% of medium mixed 10% of starting culture with a total volume of 4 litre. The air flow rate was 2.654 x 10^6 m^3. The light intensity used was 2266 lux provided by LED light. Triplicate sample were collected at 24 hr interval for 10 days. The biomass concentration was then determined by measuring optical density at 680nm using spectrophotometer.

3 Results and discussion

3.1 Design and Simulation

SolidWorks software has been used to do the analysis for the assembly, interferences, clearance, 2d drawing, and exploded view of the model. Besides, the performance of the design such as simulation also been studied and performed using the SolidWorks software. There are total 24 parts in this design. However, 19 out of 24 parts are standard which able to obtain from the market. Therefore, the dimension and shape of 19 parts are modelled according to the real dimension (Figure 1)

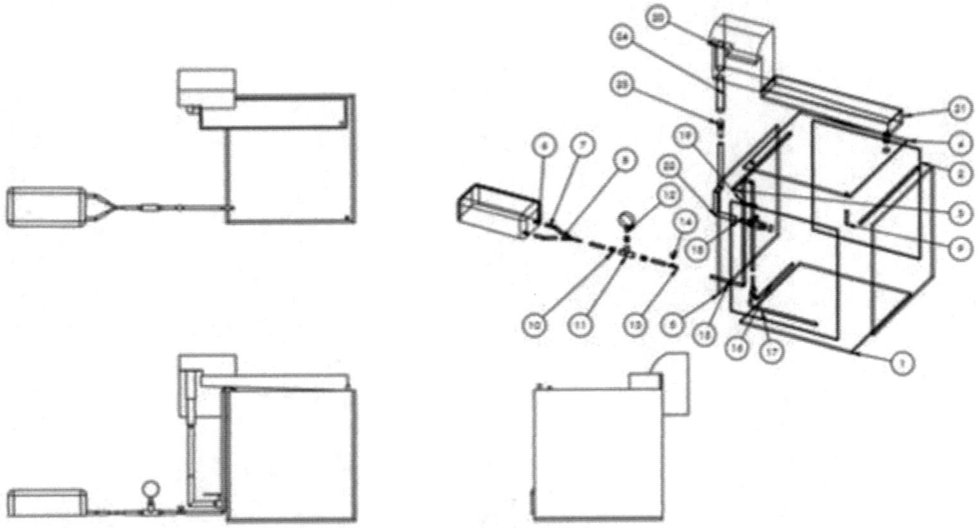

Figure 1 Drawing for final conceptual design for the tank and equipments.

During this process, the simulation of the performance of the photobioreactor is done by using some advanced simulation techniques in SolidWorks software. The simulation performed is only in the main body instead of the whole design. This is because the most important is the simulation inside the main body to show the flow inside the main body. Besides, there are two simulation that are performed for the design which one is for the air flow while another one is the water flow inside the main body (Figure 2).

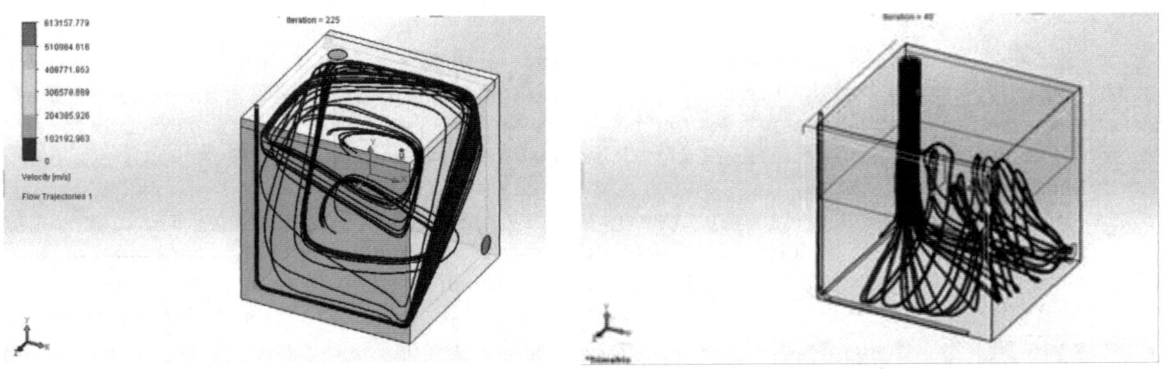

(a) (b)

Figure 2: The simulation for (a) air (b) water inside the main tank of photobioreactor

3.2 Fabrication

The material used to fabricate the main body and cover is Poly (methyl methacrylate) or more known as Perspex with 5mm thickness. Perspex is selected as the fabrication material for the product because it is transparent, and it is an important feature for cultivation of microalgae since the product need to allow light to penetrate into the main body. The 5mm thickness of Perspex sheet is used to ensure the main body is able to withstand the pressure when there is water inside and also the pressure cause by the cultivation of microalgae. After the fabrication of the main body and the procurement of components, the photobioreactor is assembled and tested the functionality as shown in Figure 3.

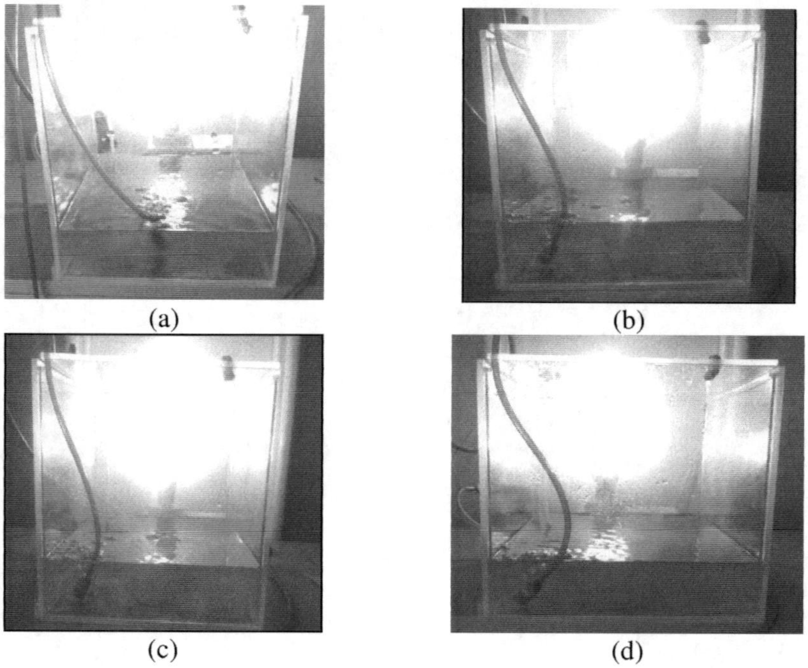

Figure 3: Cultivation of *Chlorella sp.* (a)3rd day (b) 4th day (c) 6th day (d) 9th day

3.3. Experimental analysis

In Figure 4, the growth of *Chlorella sp.* increases only until 5th day and it shows drastic decrease on 6th day. But the growth slightly increase from 7th to 10th day. Besides flocs were noticed starting from the third day in the photobioreactor. This maybe due to an excess of nutrients particularly phosphorus and nitrogen and higher concentrations of these nutrients in water cause increased growth of microalgae in the early stage of cultivation. Thus, from the it can be observed that the concentration increases rapidly from day 0 to day 1. But, the growth microalgae started to decrease from 5th day. The nutrients inside the photobioreactor is depleted as it was consumed by the rapid growing cells. At higher cell concentrations, the light intensity must be passable to penetrate through the culture. However the light intensity used for the experiment is not sufficient for the algae growth. But, too high light intensity may result in photo-inhibition and overheating environment. Thus, optimum light intensity study is required for better cell growth.

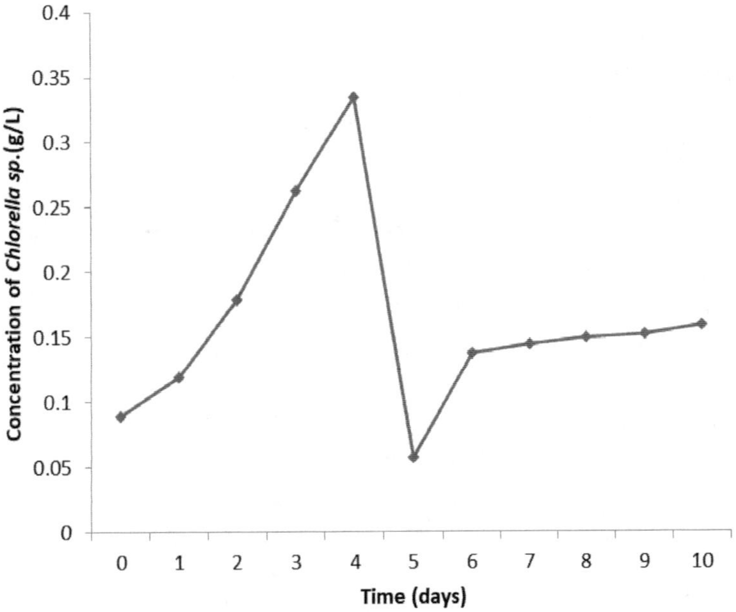

Figure 4 Concentration of *Chlorella sp.* vs time.

3.4 Comparison new photobioreactor with existing photobioreactor

According to the observation and interview session, the existing photobioreactor is only using a simple conical flask to perform the cultivation of microalgae. The existing approach was not easy to adjust light intensity because the lighting system used is a normal fluorescent bulb without a controller. Besides, the air flow rate entering the vessel is not able to accurately adjust because the vessel was not equipped with any meter or sensor to detect the air flow rate.

Therefore, the design photobioreactor provide more advantages and functions over the existing approach. The photobioreactor equipped with the pressure gauge and air flow controller to provide accurate control for air flow rates. Besides, the new photobioreactor also equipped with the lighting system that has adjustable light intensity function. The lighting system allows the changing or adjusting of light intensity with a simple switch.

4 Conclusion

The present study has been done in order to improve the design of existing photobioreactor in order to have the controllability of certain parameters (light and air/water flowrate) and improve the efficiency of photobioreactor. The new photobioreactor designed has the adjustable light intensity which is important for cultivating different species of microalgae. In addition, the photobioreactor also equipped with the accurate controllable air flow rate which allows the concentration of air/gas injected into the medium of microalgae is adjustable in order to produce optimal productivity.

5 Acknowledgement

This research is supported by the Ministry of Higher Education Malaysia and Universiti Sains Malaysia under the Fundamental Research Grant Scheme (FRGS) (Reference No.: 6071369).

6 References

[1] Patyna A and Witczak S 2016, Overview of photobioreactors for the production of biodiesel. *CHEMIK*, **70**,10, pp 634 – 643.

[2] Egbo M K, Okoani OO and Okoh IE Photobioreactors for microalgae cultivation – An overview 2017, *International Journal of Scientific & Engineering research*, vol **9** (11) pp 65-74

[3] Singh, R.N. and S. Sharma 2012, Development of suitable photobioreactor for algae production – A review. *Renewable and Sustainable Energy Reviews*, **16**(4) pp 2347-2353.

[4] Yadala, S. and S. Cremaschi 2014, Design and optimization of artificial cultivation units for algae production. *Energy*, **78** pp 23-39.

[5] Huang Q, Jiang F, Wang L and Yang C 2017, Design of Photobioreactors for Mass Cultivation of Photosynthetic Organisms. *Engineering*, **3**(3): p. 318-329.

[6] de Mooij T, de Vries G, Latsos C, Wijffels R H and Janssen M 2016. Impact of light color on photobioreactor productivity. *Algal Research*, **15**: p. 32-42.

[7] Pereira DA, Rodrigues VO, Gomez SV, Sales EA and Jorquera O 2013, Parametric sensitivity analysis for temperature control in outdoor photobioreactors. *Bioresource Technology*, **144** pp 548- 553.

[8] Maroneze M.M. Siqueira SF, Vendruscolo RG. Wagner R, de Menezes CR, Zepka LQ and Jacob-Lopes E, 2016 The role of photoperiods on photobioreactors – A potential strategy to reduce costs. *Bioresource Technology*, **219** pp 493-499.

[9] Nhat PVH, Ngo HH, Guo WS, Chang SW, Nguyen DD, Bui XT, Zhang XB and Guo JB 2018, Can algae-based technologies be an affordable green process for biofuel production and wastewater remediation? *Bioresource Technology*, **256** pp 491-501.

[10] Adeniyi, O.M., U. Azimov, and A. Burluka 2018 Algae biofuel: Current status and future applications. *Renewable and Sustainable Energy Reviews*, **90** pp 316-335.

IOP Conf. Series: Materials Science and Engineering **716** (2020) 012011 doi:10.1088/1757-899X/716/1/012011

Isolation and characterization of novel acidophilic microalgae from abandoned mining site area for carbohydrate biosynthesis and its kinetic growth study in photobioreactor

K M Tan*[1], M A Kassim*[1], Z J Ng[1] and J Lalung[1]

[1]Bioprocess Technology Division, School of Industrial Technology, University Sains Malaysia, 11800 Penang, Malaysia.

*Email: km_tan10@hotmail.com; asyrafkassim@usm.my

Abstract. Microalgae have gained attention as an alternative feedstock for biofuel production due to its potential to accumulate carbohydrate, protein and lipid. However, current microalgae culture technology has been impeded by low biomass production, carbohydrate accumulation and contamination issue. In order to overcome this limitation, this study involves the isolation, screening and identification of acidophilic microalgae from Frog Hill. Then, the effect of different initial pH on growth and carbohydrate productivity was studied toward the selected microalgae. Further cultivation was done in indoor and outdoor using 15L fabricated PBR under optimize pH condition. At the end of experiment, the growth kinetic and carbohydrate productivity for both conditions were evaluated. Total of 4 native acidophilic microalgae were successfully isolated and was identified using 18s RNA. The *Scenedesmus parvus* was exhibited highest biomass production and carbohydrate productivity up to 1.2640 ± 0.001 g L^{-1} and 15.47 ± 0.048 mg L^{-1} d^{-1} under highly acidic condition. It also showed a better performance under outdoor cultivation up to 9.54% compared to indoor cultivation. Hence, this research paper showed that the potential of novel acidophilic microalgae survives under extremely condition with high carbohydrate biosynthesis, which could be useful for industrial application.

1. Introduction

The rapid growing of world population required continuously increase the global demand and fuel energy. Some of the researchers also predicted that world fossil oil reserves will be exhausted in less than 50 years due to high consumption rate or demand by mankind [1]. Therefore, the renewable energy resource plays an important role to replace the exhausted fossil fuel. Microalgae becomes a promising feedstock for biofuel production due to the number of advantages, including shorter life cycle compared to terrestrial plant, carbon dioxide (CO_2) bio-sequestration capability, not competing for arable lands, and accumulate chemical metabolites such as lipid, protein and carbohydrate [2]. Microalgal-carbohydrate is one of the dominant compound in microalgae biomass. It can act as the valuable chemical platform to be converted into biofuel, bioethanol, and biopolymer through biotechnological application using sustainable manner [3]. Previous study showed that *Chlorococum sp.* microalgae biomass can be used as feedstock for bioethanol production up to 3.83 g L^{-1} fermented using *Saccharomyces bayanus* under optimum condition [4].

However, the industrial production of microalgae-based products are not yet economically feasible and not cost-effective. This situation normally due the low microalgae biomass production. One of these factor was contributed by serious contamination by unwanted microorganism which can impede the microalgae growth. Hsu and Wu [5] proved that using large scale cultivation could be resulting in lower yield due to many unknown environmental factors that can directly affect the microalgae cultivation process. Besides that, Hirooka et al. [6] study also shows that mass cultivation using open pond system was easily contaminated by undesired microorganism in which suppressing the microalgae growth and resulting in the low productivity. Then, this problem has led to increase in total cost of cultivation and resulting in difficulties of downstream purification process.

Hence, an innovation idea of large scale cultivation of microalgae using closed photobioreactor system under controlled conditions have been suggested by previous studies in which could increase the biomass productivity to ensure the feasibility of product formation [7]. Similar study also be conducted by Sánchez Mirón et al. [8] who demostrates that *Phaeodactylum tricornutum* microalgae was achieved biomass concentration up to 4 kg m^{-3} cultivated using 60 L of bubble column, split-cylinder airlift device and concentric draft-tube sparged airlift vessel respectively. This would be beneficial for additional product value such as omega-3 polyunsaturated fatty acids production. Besides that, Schenk et al. [9] have suggested that cultivating extremophile microalgae that can tolerate at particular condition that is hostile for other contaminants, at the meanwhile could enhance the biomass production.

Up to date, there was still scarce of information regarding the acidophilic microalgae cultivated under extreme condition for carbohydrate polymer biosynthesis. Thus, the main objective of this study was to isolate the most tolerant acidophilic microalgae and determine the carbohydrate polymer biosynthesis cultivated under acidic condition. At the end of the experiment, the suitable kinetic model was chosen based on isolated microalgae growth rate cultivated in large scale of fabricated tubular photobioreactor.

2. Materials and Methods

2.1 Sample collection and isolation of acidophilic microalgae

The small portion of soil containing green microalgae was collected from the abandoned mining site area, Frog Hill (5.4354° N, 100.4847° E). Then, the sample containing microalgae were enriched in autoclaved acidic bold basal medium, BBM (pH 3.0) in 250mL conical flask, sparging using 15% (v v^{-1}) CO_2 concentration. The chemical composition of BBM as followed: $NaNO_3$ 25 g L^{-1}, $MgSO_4.7H_2O$ 7.5 g L^{-1}, NaCl 2.5g L^{-1}, K_2HPO_4 7.5 g L^{-1}, KH_2PO_4 17.5 g L^{-1}, $CaCl_2.2H_2O$ 2.5 g L^{-1}, $ZnSO_4.7H_2O$ 8.82 g L^{-1}, $MnCl_2.4H_2O$ 1.44 g L^{-1}, MoO_3 0.71 g L^{-1}, $CuSO_4.5H_2O$ 1.57 g L^{-1}, $Co(NO_3)_2.6H_2O$ 0.49 g L^{-1}, H_3BO_3 11.42 g L^{-1}, EDTA 50 g L^{-1}, KOH 31 g L^{-1}, $FeSO_4.7H_2O$ 4.98 g L^{-1} and H_2SO_4 1 mL.

Next, the sample was isolated by combination method between dilution with agar streaking technique for obtaining the axenic culture. Then the single colony was inoculated into BBM medium under controlled culture room at 30 ± 2°C, illuminated light with photon intensity of 20.99 μmol m^{-1} s^{-1} provided by cool daylight tube with photoperiod cycle of 12 h: 12 h (light: dark).

2.2 Experimental set up

A total volume of 70 mL with OD 1.00 (10% v v^{-1}) microalgae cell was inoculated into 630 mL BBM medium similar with above conditions. Then, the cell growth rate was measured in daily basis from day 0 until days 15 and the triplicate readings were taken. The microalgal cells were recovered by centrifugation and dried using oven at 60 °C. The dried microalgal pellet was subjected for further analysis. The selected strain that possess highest growth rate with carbohydrate productivity will be subjected to different pH values medium ranging from pH 3.00 to pH 11.00 in order to study the kinetic behavior and carbohydrate accumulation. The specific growth rate (d^{-1}) and carbohydrate productivity (mg L^{-1} d^{-1}) were calculated according to the following equation (1) and (2):

$$k = \frac{\ln N_1 - \ln N_2}{t_1 - t_2} \quad (1)$$

where k (d^{-1}) is the specific growth rate, N_1 and N_1 are the biomass concentration at day t_1 and t_2

$$CP = BP \times CW \tag{2}$$

where CP is carbohydrate productivity (mg L^{-1} d^{-1}), CW (%) is carbohydrate content (w w^{-1}) based on dry weight.

2.3 Measurement and quantification of total carbohydrate and sugar component

The total carbohydrate content was determined using phenol-sulfuric acid method following the method by Nielsen [10]. Dextrorotatory (D) (+) glucose was used as a standard for this carbohydrate analysis The microalgae that contain highest carbohydrate productivity will be selected for the subsequent experiment.

2.4 Morphological identification of microalgae

Identification of the microalgae species was done through conventional method and molecular approach. Isolated strains were preliminary identified using standard morphological features. Subsequently, the total genomic DNA from the cultured sample was extracted using GF-1 plant DNA extraction kit (Vivantis Technologies, Malaysia) according to the manufactor instruction. For PCR amplification, the extracted DNA was used to amplify the 18S rRNA with two general primers, forward primer (EUF) and reverse primer (EUR). The PCR analysis was performed by Mastercycler EP384 Thermal Cycler (Eppendorf, US) with the conditions described by Nübel et al. [11].

EUF: 5'-GTCAGAGGTGAAATTCTTGGATTTA-3'
EUR: 5'-AGGGCAGGGACGTAATCAACG-3'

The PCR product was further analysed by 18S rRNA gene sequencing at Centre for Chemical Biology, CCB Penang using EUF and EUR primers. All sequences obtained here were blasted through NCBI GeneBank database and performed phylogenetic tree analysis.

2.5 Phylogenetic analysis

A search of NCBI GenBank with BLAST was used to identify the isolated microalgae species with rRNA gene sequences downloaded from the database. The sequences containing fewer than 200 nucleotides or in excess of 1000 nucleotides were removed, and sequences not belonging under microalgae species also be excluded from this study. Subsequently, all the downloaded sequences were aligned accordingly using the ClusterX 2.1 software for restricting the multiple sequences alignment within the minimum common region. The phylogenetic tree was constructed using the neighbor-joining (NJ) method and Kimura's two-parameter model implemented within Mega-X software.

2.6 Kinetic growth in photobioreactor

A 10% (v/v) of selected microalgae strain was further up-scaled into 15L of fabricated tubular PBR with optimum pH condition cultivated under indoor and outdoor environment. The cell growth and carbohydrate productivity were also evaluated under both cultivation condition. The basic Logistic equation and Gompertz equation from Sigmaplot version 12.5 as in equation (3) and (4) were adopted in this research to predict the kinetic behavior of microalgal growth cultivated under PBR system. Choosing a suitable model with $R^2 > 0.95$ showed that the kinetic growth of microalgae fitted well with the suggested model.

Logistic model:
$$C(t) = C_o e^{\mu t} \times [1 - (\frac{C_o}{C_m})(1 - e^{\mu t})]^{-1} \tag{3}$$

Gompertz model:
$$C(t) = C_0 \times e[-e(\mu \times e^{1/C_0})] \times (C_m - t) + 1 \tag{4}$$

where Co is the initial microalgae biomass (g L^{-1}); Cm is the maximum biomass (g L^{-1}); μ is the specific growth rate (day^{-1}) and t is the cultivation time (day).

2.7 Statistical analysis
All results were expressed in triplicate to calculate the mean ± standard error. A value of $P<0.05$ was performed by using Minitab 16 software in order to consider the results were statistically significant.

3. Results and Discussions
3.1 Selection and identification of the most promising strain among the candidate strains
A total number of 4 microalgae strains namely USM_S1, USM_S2, USM_S3 and USM_S4 were successfully isolated from the soil sample. Further screening on microalgae capability to exhibit high growth rate and carbohydrate productivity under extreme conditions (low pH and high CO_2 concentration) were carried out and the result was tabulated in Table 1.

Result showed that the USM_S1 has a great deal in their ability to produce highest cell dry weight, biomass productivity and specific growth rate which were 1.2640 ± 0.001 g L^{-1}, 0.078 ± 0.004 g L^{-1} d^{-1} and 0.1691 ± 0.001 d^{-1} respectively under extreme condition. This result was obviously showed that strain 1 grew profusely compared to other isolates under favorable condition. Similar trend was observed by Vaquero et al. [12] who shows that *Coccomyxa onubensis* exhibits highest specific growth rate under acidic and high CO_2 condition. This result was further agreed by the previous study who demonstrated that every microalgae species could exhibit different growth rate under certain environment condition [1]. Interestingly, this USM_S1 also accumulates a moderately high amount of carbohydrate up to 37.19 ± 0.160% with highest in carbohydrate productivity 15.47 ± 0.048 mg L^{-1} d^{-1} under the same cultivation condition (Table 1). Hence, it can be concluded that this specific strain of microalgae was a good candidate to be served as carbohydrate-based feedstock for biofuel and chemicals production. The strain that possess the highest growth rate and carbohydrate productivity (USM_S1) was further subjected for molecular identification and large scale mass cultivation.

Table 1. Kinetic growth rate and carbohydrate productivity of microalgae cultivated under acidic condition (pH 3.00 and 15% CO_2 concentration)

Microalgae	Maximum cell dry weight (g L^{-1})	Biomass Productivity (g L^{-1} d^{-1})	Specific growth rate, μ (d^{-1})	Carbohydrate content (%) (w w^{-1})	Carbohydrate productivity (mg L^{-1} d^{-1})
USM_S1	1.2640 ± 0.001	0.078 ± 0.004	0.1691 ± 0.001	37.19 ± 0.160	15.47 ± 0.048
USM_S2	0.5623 ± 0.001	0.036 ± 0.001	0.1110 ± 0.001	22.17 ± 1.036	7.98 ± 0.037
USM_S3	0.1792 ± 0.001	0.007 ± 0.001	0.0190 ± 0.001	13.12 ± 0.618	0.92 ± 0.004
USM_S4	0.4902 ± 0.001	0.028 ± 0.001	0.1042 ± 0.001	22.91 ± 0.186	6.42 ± 0.005

The identification pf the selected microalgae strain (USM_S1) was projected through morphological features and 18S rRNA sequence with the help of universal primer (EUF and EUF). Figure 1 (a) shows the morphology features of USM_S1 under 100x oil immersion microscopic examination. Based on the figure, it shows that USM_S1 was unicellular microorganism and appeared in form colonies. The structure of the cells was ovoid cylindrical with the dimension of 3 μm broad and 4 μm long. This preliminary result showed that USM_S1 was closely related to *S. parvus* as described by Reddy and Chaturvedi [13].

Further identification work was carried out through the 18S rRNA sequence analysis. The NCBI BLAST search was performed using the sequence as a query showed that it is most closely related to FR865718.1 *S. parvus* with 99% similarity. This strain was further confirmed through the multiple sequence alignment and phylogenetic tree analysis. The result appeared was coincided with the previous result, which was *S. parvus* (Figure 1(b)).

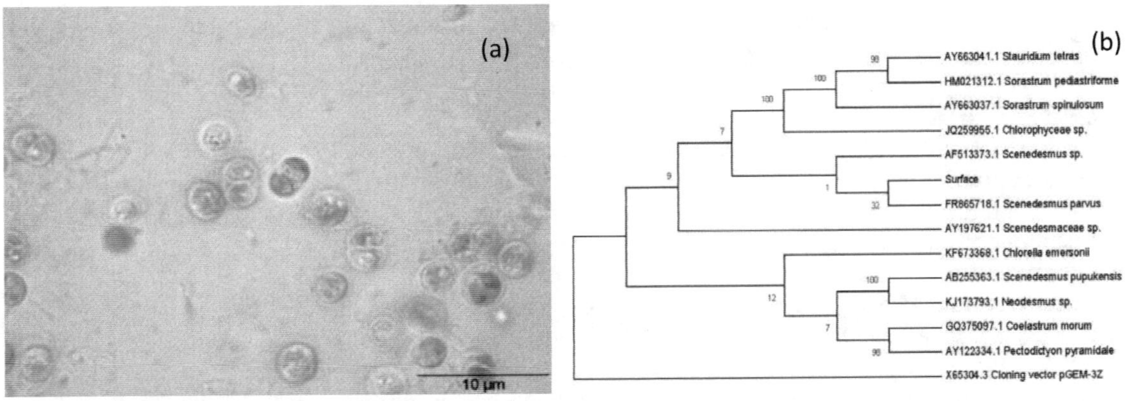

Figure 1. (a) Morphological identification of *S. parvus* under 100 X magnification oil immersion light microscope. (b) Phylogenetic trees analysis of locally isolated microalgae strain with most similar sequences retrieved from NCBI GenBank database.

3.2 Effect of different initial pH on S. parvus growth and carbohydrate productivity

The isolated microalgae was further tested in different pH cultivation medium ranging from 3.00 to pH 11.00 in order to study the extent of pH effect on *S. parvus* growth. Based on Figure 2 (a), it shows that *S. parvus* obtained highest biomass production up to 1.339 ± 0.001 g L^{-1} under acidic condition (pH 3.00) compared to alkaline condition (pH 11.00) which showed the lowest cell dry weight 0.5178 ± 0.007 g L^{-1} over 15 days of cultivation time. Different result has been observed by Ren *et al.*[14], who displayed a lower growth rate of *Scenedesmus sp.* strain R-16 when cultivated under acidic condition, pH 3.0. Whereas, Cortés et al. [15] shows that wild-type *Dictyosphaerium chlorelloides* exhibited a maximum growth rate under alkaline condition, pH 9.0 This clearly indicated that the capability of microalgae to tolerate with different pH was species-specific. Varying the initial pH of the cultivation medium also could directly affect the metabolic process in microalgae cells [16].

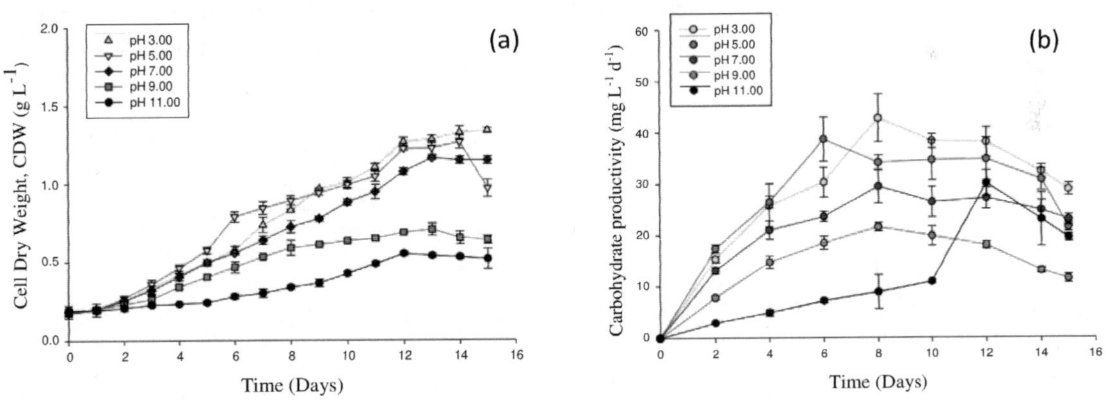

Figure 2. (a) Cell dry weight and (b) carbohydrate productivity of *S. parvus* growth over 15 days of cultivation under acidic condition (pH 3.00).

The Figure 2 (b) showed the highest carbohydrate productivity was found 42.76 mg L^{-1} d^{-1} at day when the cultivation medium was shifted to acidic condition (pH 3.00). This was due to the stress condition applied and causing the changing in metabolism of microalgae and subsequently enhance the carbohydrate productivity [2]. Al-Safaar *et al.*[17] also demonstrates that increased in carbohydrate content of *Chlorella vulgaris* and *Chroococcus minor* when applied under acidic condition. Hence, it was clearly illustrated that by adjusting the pH cultivation medium could directly affect the microalgal growth rate and carbohydrate productivity.

3.3 The biomass production and carbohydrate productivity of S. parvus cultivated under indoor and outdoor condition using 15 L PBR

In order to ensure the economic feasibility of the biofuel production, therefore, selection of fast-growing microalgae strain with high carbohydrate accumulation is necessary to develop in pilot-scale cultivation to achieve and maintain the year-round harvesting capability [18]. Hence, this context has involved in identification of the selected strain to achieved in maximum cell growth rate and carbohydrate productivity by applying large scale microalgae cultivation up to 15 L closed system PBR.

S. parvus (Strain 1) was selected due to the highest cell density and carbohydrate productivity cultivated under highly acidic condition (Table 1). Further up-scaled cultivation was done of *S. parvus* towards two different cultivation conditions including indoor and outdoor in order to study the kinetic behavior of this selected strain. Figure 3 (a) showed that outdoor cultivation of *S. parvus* displayed a better growth result up to 0.9326 ± 0.083 g L^{-1} compared to indoor cultivation which only exhibited 0.8514 ± 0.074 g L^{-1} at days 9 of cultivation. The difference in cultivation growth result was due to the cultivation conditions which including temperature and light intensity. During the outdoor cultivation, higher temperature (33.3 ± 1.74 °C) and light intensity (20648.33 ± 20025.90 lux) was achieved by *S. parvus* over the cultivation period (result not shown). These would favor microalgae photosynthesis rate and subsequently resulting in higher growth rate [19].

Different result has been obtained by Ritcharoen et al. [20] who showed that indoor cultivation of *Chaetoceros gracilis* was performed better compared to outdoor cultivation using flat-panel airlift photobioreactor. This phenomenon was due to the uneven light availability received during outdoor cultivation and resulting in low microalgae growth rate. Nonetheless, the feasibility of outdoor cultivation was further been agreed by previous study, Wen et al. [21] who showed the outdoor cultivation using pilot-scale raceway have a nice adaptability for *Graesiella sp.* WBG-1 compared to indoor experiment. Similar result has been observed by previous study who showed the biomass productivity of *Skeletonema costatum* microalgae was also higher up to 28.0 ± 0.16 mg L^{-1} d^{-1} compared to indoor cultivation [22]. Although outdoor cultivation of microalgae has encountered strong fluctuation of natural solar irradiance and temperature, however, these specific microalgae still able to perform well under harsh condition. Hence, this showed that *S. parvus* is one of the most tolerance strain that could sustain in wide range of temperature and light intensity, which could be applied for outdoor pilot-scale cultivation.

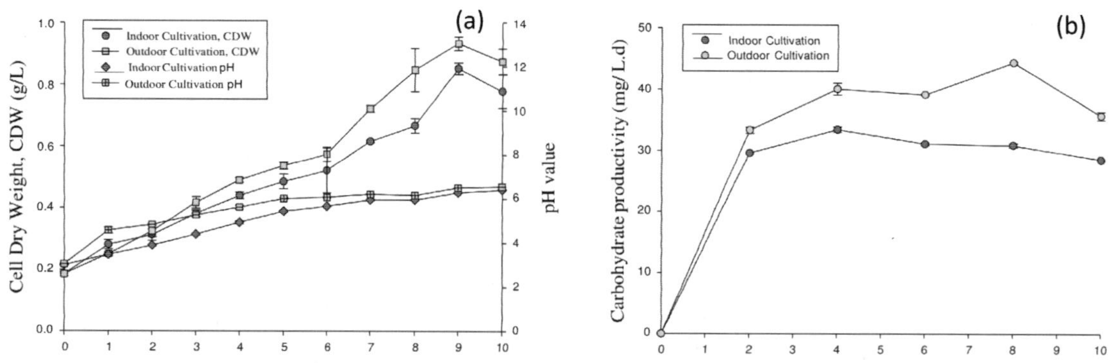

Figure 3. (a) Cell dry weight (g L^{-1}) and (b) Carbohydrate productivity (mg L^{-1} d^{-1}) of *S. parvus* in 15 L closed photobioreactor over 10 days of cultivation time under acidic condition (pH 3.00)

Carbohydrate productivity of *S. parvus* under indoor and outdoor cultivation was also evaluated in this study. Based on Figure 3 (b), it showed that highest carbohydrate productivity could be achieved by *S. parvus* up to 44.35 ± 0.1693 mg L^{-1} d^{-1} (days 8) under acidic outdoor cultivation. This value was higher up to 32.74% compared to the indoor cultivation. This was due to the changes in environmental cultivation condition and subsequently affect the carbohydrate accumulation. This result was coinciding with the previous study who figured out the increasing in temperature and irradiance could enhance the carbohydrate productivity up to 111.8 mg L^{-1} d^{-1}, which was 2.5% higher during sunny environment compared to rainy and cold condition, in Taiwan [23]. This was most probably due to biological processes within microalgae in which can contribute to turnover of enzyme, metabolic activity, growth rate and carbohydrate accumulation with the respect of temperature surrounding [23, 24]. Other study was done by Show *et al.*, [25] who also reported that at low temperature could affects the viscosity of cytoplasm, leading to inefficient of nutrient utilization, or resulting in photo-inhibition. Hence, the photosynthesis mechanism of microalgae would become not effective and subsequently lower the cell growth rate. Therefore, by selecting the robust microalgae was important in this context in order to have capability to tolerate with high temperature range and accumulate the significant amount of carbohydrate.

3.4 Kinetic growth model study of microalgae

The growth kinetic study should be applied to achieve a higher understanding of microalgae growth under different cultivation conditions. For this particular study, two kinetic growth models namely Logistic and Gompertz model were used to compare the suitable models that can fit well with the experimental microalgae growth data. Result shows that cultivation under indoor condition attained a higher R^2 value of 0.9678 in Gompertz model, compared to 0.9677 in Logistic-type model (Table 2). This mean Gompertz model fit well with the experiment growth data. This result was further confirmed by the RMSD result shown in Table 2. A similar result was obtained in which Gompertz model shows lower RMSD compared to Logistic model. RMSD is represented a good fit to extent the correctness and the lower value specifies the finest model agreement [26]. Hence, it shows that Gompertz model is acceptable to represent the growth of *S. parvus* under indoor cultivation due to lower RMSE value and vice versa for the outdoor cultivation.

Table 2. Performance indicates of the nonlinear models of *S. parvus* growth under different types of cultivation

Types of cultivation	Cell growth model			
	Logistic		Gompertz	
	R^2	RMSD	R^2	RMSD[a]
Indoor	0.9677	0.0378	0.9678	0.0377
Outdoor	0.9771	0.0390	0.9760	0.0399

[a]RMSD: Root Mean Square Deviation

The suitability of Gompertz growth model was further studied by Praveen et al. [27], who showed that Gompertz model was suitable models (high R^2 and low RMSE values) for describing the growth kinetic of *Chlorella sp.* under photoautotrophic, heterotrophic and photomixotrophic conditions. Other study also indicated that high R^2 value which denoted by the logistic-type model showed that the suitability of this model for microalgae growth under normal condition and nitrogen-depletion condition [28]. Hence, it can be concluded that the suitability of the model was largely contributed by the differences of microalgae strains, growth medium, cultivation conditions and mode of cultivation [29]. By choosing the right model is important for predicting the microalgae growth characteristics and maximum biomass production under specific cultivation condition.

4. Conclusions

This present study revealed that the *S. parvus* able to achieve highest biomass prouction and carbohydrate productivity under extreme environment which were 1.2640 ± 0.001 g L^{-1} and 15.47 ± 0.048 mg L^{-1} d^{-1} respectively. Then, in order to ensure the economical feasibility of the carbohydrate biosynthesis polymer, the isolated *S. parvus* was further cultivated in large scale production. The results

IOP Conf. Series: Materials Science and Engineering **716** (2020) 012011 doi:10.1088/1757-899X/716/1/012011

indicated that isolated acidophilic microalgae were suitable for carbohydrate polymer biosynthesis up to 44.35 ± 0.1693 mg L^{-1} d^{-1}. This is important to ensure the economic feasibility of next generation biofuel production. Lastly, this study annotates that the model is condition dependent in which different cultivation condition suitable different kinetic model. This information is useful for predicting the microalgae growth characteristics and can act as an essential platform when the cultivation is scaled-up to commercial application.

.

5. References

[1] Abou-Shanab R A, Hwang J H, Cho Y, Min B and Jeon B H 2011 Characterization of microalgal species isolated from fresh water bodies as a potential source for biodiesel production *Applied Energy* **88** 3300-06

[2] Chen C Y, Zhao X Q, Yen H W, Ho S H, Cheng C L, Lee D J, Bai F W and Chang J S 2013 Microalgae-based carbohydrates for biofuel production. *Biochemical Engineering Journal* **78** 1-10

[3] Kassim M A, Rashid M A and Halim R 2017 Towards biorefinery production of microalgal biofuels and bioproducts: production of acetic acid from the fermentation of *Chlorella sp*. and *Tetraselmis suecica* hydrolysates. *Green and Sustainable Chemistry* **7** 152-71

[4] Harun R, Danquah M K and Forde G M 2010 Microalgal biomass as a fermentation feedstock for bioethanol production *Journal of Chemical Technology & Biotechnology* **85** 199-203

[5] Hsu Y L and Wu W T 2002 A novel approach for scaling-up a fermentation system. *Biochemical Engineering Journal* **11** 123-30

[6] Hirooka S, Higuchi S, Uzuka A, Nozaki H and Miyagishima S Y 2014 Acidophilic green alga *Pseudochlorella* sp. YKT1 accumulates high amount of lipid droplets under a nitrogen-depleted condition at a low-pH *PloS One* **9** e107702 [5] Mallick N, Bagchi S K, Koley S and Singh A K 2016 Progress and challenges in microalgal biodiesel production. *Frontiers in Microbiology* **7** 1019-30

[7] Mallick N, Bagchi S K, Koley S and Singh A K 2016 Progress and challenges in microalgal biodiesel production. *Frontiers in Microbiology* **7** 1019-30

[8] Sánchez Mirón A, Garcia Camacho F, Contreras Gomez A, Grima E M and Chisti Y 2000 Bubble-column and airlift photobioreactors for algal culture *AIChE Journal* **46** 1872-87

[9] Schenk P M, Thomas-Hall S R, Stephens E, Marx U C, Mussgnug J H, Posten C, Kruse O and Hankamer B 2008 Second generation biofuels: high-efficiency microalgae for biodiesel production *Bioenergy Research* **1** 20-43

[10] Nielsen S S 2010 Phenol-sulfuric acid method for total carbohydrates *Springer: Food analysis laboratory manual* 47-53

[11] Nübel U, Garcia-Pichel F and Muyzer G 1997 PCR primers to amplify 16S rRNA genes from cyanobacteria *Appl. Environ. Microbiol.* **63** 3327-32

[12] Vaquero I, Vázquez M, Ruiz-Domínguez M and Vílchez C 2014 Enhanced production of a lutein-rich acidic environment microalga *Journal of Applied Microbiology* **116** 839-50

[13] Reddy B M and Chaturvedi A 2015 *Scenedesmus* and related genera in riverine system of Chandrapur district, Maharashtra, India *Phykos* **45** 19-28

[14] Ren H Y, Liu B F, Ma C, Zhao L and Ren N Q 2013 A new lipid-rich microalga *Scenedesmus sp.* strain R-16 isolated using Nile red staining: effects of carbon and nitrogen sources and initial pH on the biomass and lipid production *Biotechnology for Biofuels* **6** 143-53

[15] Cortés A A, Sánchez-Fortún S, García M and Bartolomé M C 2018 Effects of pH on the growth rate exhibited of the wild-type and Cd-resistant *Dictyosphaerium chlorelloides* strains *Limnetica* 37 229-38

[16] Kosourov S, Seibert M and Ghirardi M L 2003 Effects of extracellular pH on the metabolic pathways in sulfur-deprived, H_2-producing *Chlamydomonas reinhardtii* cultures. *Plant and Cell Physiology* **44** 146-55

[17] Al-Safaar A T, Al-Rubiaee G H and Salman S K 2016 Effect of pH condition on the growth and lipid content of microalgae *Chlorella vulgaris* & *Chroococcus minor Journal of Scientific & Engineering Research* **7** 1138-43

[18] John R P, Anisha G, Nampoothiri K M and Pandey A 2011 Micro and macroalgal biomass: a renewable source for bioethanol *Bioresource Technology* **102** 186-193

[19] Hancke K, Hancke T B, Olsen L M, Johnsen G and Glud R N 2008 Temperature effects on microalgal photosynthesis-light responses measured by O_2 production, pulse-amplitude-modulated fluorescence, and 14C assimilation 1 *Journal of Phycology* **44** 501-514

[20] Ritcharoen W, Sriouam P, Nakseedee P, Sang P, Powtongsook S, Kungvansaichol K and Pavasant P 2014 Cultivation options for indoor and outdoor growth of *Chaetoceros gracilis* with airlift photobioreactors *Maejo International Journal of Science and Technology* **8** 100-113

[21] Wen X, Du K, Wang Z, Peng X, Luo L, Tao H, Xu Y, Zhang D, Geng Y and Li Y 2016 Effective cultivation of microalgae for biofuel production: a pilot-scale evaluation of a novel oleaginous microalga *Graesiella sp.* WBG-1 *Biotechnology for Biofuels* **9** 123-35.

[22] Pérez L, Salgueiro J L, González J, Parralejo A I, Maceiras R and Cancela Á 2017 Scaled up from indoor to outdoor cultures of *Chaetoceros gracilis* and *Skeletonema costatum* microalgae for biomass and oil production *Biochemical Engineering Journal* **127** 180-187

[23] Ho S H, Chen Y D, Chang C Y, Lai Y Y, Chen C Y, Kondo A, Ren N Q and Chang J S 2017 Feasibility of CO_2 mitigation and carbohydrate production by microalga *Scenedesmus obliquus* CNW-N used for bioethanol fermentation under outdoor conditions: effects of seasonal changes *Biotechnology for Biofuels* **10** 27-40

[24] Franco M C, Buffing M F, Janssen M, Lobato C V and Wijffels R H 2012 Performance of *Chlorella sorokiniana* under simulated extreme winter conditions *Journal of Applied Phycology* **24** 693-699

[25] Show P, Tang M, Nagarajan D, Ling T, Ooi C W and Chang J S 2017 A holistic approach to managing microalgae for biofuel applications *Int. J. of Mol. Sci.* **18** 215-49

[26] Nigam S, Rai M P and Sharma R 2011 Effect of nitrogen on growth and lipid content of *Chlorella pyrenoidosa Am J Biochem Biotechnol* **7** 124-129

[27] Praveen K, Abinandan S, Natarajan R and Kavitha M 2018 Biochemical responses from biomass of isolated *Chlorella sp.*, under different cultivation modes: non-linear modelling of growth kinetics *Brazilian Journal of Chemical Engineering* **35** 489-96

[28] Surendhiran D, Vijay M, Sivaprakash B and Sirajunnisa A 2015 Kinetic modeling of microalgal growth and lipid synthesis for biodiesel production *3 Biotech* **5** 663-69

[29] Lammers P J *et al.*2017. Review of the cultivation program within the National Alliance for Advanced Biofuels and Bioproducts *Algal research* **22** 166-86

Acknowledgements

The authors wish to thank School of Industrial Technology, Universiti Sains Malaysia and the Ministry of Higher Education, Malaysia (USM). The project and manuscript preparation were financially supported by the RUI Grant 1001/PTEKIND/8011043 and USM Fellowship RU(1001/CIPS/AUPE001) from Universiti Sains Malaysia

IOP Conf. Series: Materials Science and Engineering **716** (2020) 012012 doi:10.1088/1757-899X/716/1/012012

EXTRACTION AND CHARACTERIZATION OF STARCH FROM MICROALGAE AND COMPARISON WITH COMMERCIAL CORN STARCH

R N Ramli.[1],*C K Lee[1], M A Kassim[1]

[1]Bioprocess Technology Division, School of Industrial Technology, UniversitiSains Malaysia, 11800 USM, Pulau Pinang, Malaysia.

*Email: rozinuraika@gmail.com ; cklee@usm.my

Abstract. Starch is the main component that consumed one third in the human diet. One-third of starch produced in the world is used in non-food application such as rubbers production and plastic. Plastic produced from starch which called as bio-plastic is biodegradable and helps in reducing the world pollution which non-degradable plastic now endanger the ecosystem either land or water. Microalgal the easy growing and environmental friendly have opened new potential of revolution for starch-based bio-plastic. The small granules of starch produced from microalgae possess an advantage to replace the starch from food-based, as the small granules able to yield a good quality of bio-plastic. In this study, marine microalgae had been studied for its starch content and could reach up to 19% of total starch content. However, in order to ensure the starch produced from marine microalgae is suitable for bio-plastic application, the characteristics of starch had been studied based on the size of starch granules, amylose/amylopectine content, swelling power and solubility and turbidity. These analyses were done with a comparison of corn starch that used widely in the production of bio-plastic. Hence, by determining these crucial characteristics, the application of starch from microalgae could be expended to various sustainable products.

1 Introduction

Starch is a natural carbohydrate storage polymer accumulated in plants. It is a completely biodegradable polysaccharide and one of the most abundant renewable resources. It has been considered as an excellent candidate to partially substitute synthetic polymer in packaging and other low-cost applications due to its abundance, biodegradability and low cost [1]. The molecular structure that consist of two D-glucose polymers: amylose and amylopectin. Amylose is a linear structure with α-1,4-linked glucan, and amylopection is branched molecule characterized by α- 1,4- linked and α- 1,6 branching links [2].

The usage of starch in daily life is well-known either in food or non-food product. The sources of starch are from crops as potato, corn, and tapioca. Gifuni *et al.* [3] wrote in the study, two third of the starch used in food and one third in non-food.

The awareness on sustainable issue had hit the market as now starch crops had become the feedstock for the green product such as bioethanol and bioplastic. United State (U.S) on 2009 is the top producer for bioethanol production using corn as feedstock as constitute 50.7% from all world market while Brazil as the second producer, 40.2% production in market by using sugar cane as main sources [4]. These two sources are good in abundance and renewable source. However, these sources main is food sources which will lead to crisis in increasing of food price and also will create an issue between food sources and non-food production. In 2005, U.S was able to produce only 18.9 billion L of bioethanol from 20% of corn land in U.S. This production was counted only 1% from total fuel consumed in U.S per year [4]. Hence, in order to increase more the yield, demanding on these crops will increase as per the usage of land will be wider and will lead to deforestation. The process to produce these crops will never be sustainable even the product yield is green renewable product.

The chances of different starch properties will also affect due to the physical factors as the weather changes and uncontrollable environment conditions. The component inside the plant plays big role in product formation. Starch characteristics are determined by the granule size, amylase/amylopectin ratio, turbidity and solubility and swelling power [3].

These issues had lead to propose to use microalgae as the excellent candidates for starch accumulation and production. The advantages of being fast growing, easily grow and the capability absorbing more carbon dioxide and greenhouse emitting for photosynthesis [5].The other prominent features of microalgae is it has simple growth requirement, can be easily grown in various aquatic environments such as fresh water, saline water or municipal waste water and also in controlled condition which will lead to consistency of cell produce without affecting from the outer environment condition [6,7]. Starch contents in microalgae species will be based on cultivation conditions and cultivation time. Generally, total carbohydrate content in microalgae is about 20% dry weight (DW) and starch content is about 10% DW [8]. The low-cost yet high yield from microalgae had reported by Brányiková *et al.*[9] by using freshwater alga *Chlorella vulgaris* which able to produce almost 60% of starch content since it accumulates large amounts of carbohydrates through photosynthesis and stores as starch in cells.

This study was to compare the characteristic of starch extracted from microalgae with corn starch in order to compare the ability of starch microalgae could be substituent to the starch from crop. The corn starch used in this study was commercial fully purified corn starch.

2 Methodology

2.1 Microorganism

*Klebsormidium flaccidum*was isolated from sea in Penang Island, Northern of Malaysia which further isolated in Bioprocess Division laboratory cell culture room. The cell was isolated using method of single-cell isolation by micropipette and dilution technique [10]. The medium used was Walne's medium (g/L) at 35 ppt: {Solution A: $FeCl_3.6H_2O$,1.5; $MnCl_2.4H_2O$, 0.4; H_3BO_3, 33.6; Na_2-EDTA, 45; $NaH_2PO_4.2H_2O$, 20; $NaNO_3$, 100, Solution B; 1 mL [in 100 mL: $ZnCl_2$, 2.1g, $CoCl_2.6H_2O$, 2.0g, $(NH_4)_6Mo_7O_{24}.4H_2O$, 0.9g, $CuSo_4.5H_2O$, 2.0g, Concentrated HCl, 10mL], The strain was maintained in Walne's media at 35 ppt in room temperature (26°C-30°C). Solution C: [in 200 mL, Thiamine HCl (vitamin B_1), 0.2 g, Cobalamin (vitamin B_{12}), 10 mg.] Solution D: (in 1 liter) $Na_2SiO_3.5H_2O$, 40.0g}. To complete the media, mix Solution A for 1.0 mL, Solution B for 0.001 mL, Solution C 0.1mL and Solution D 2.0 mL in per liter of seawater at pH 7.5. Solution D was added for isolation diatoms only [11]. The cultivation of cell was done in photobioreactor (Infors HT) at 12:12 hours of light and dark, at 125 rpm, temperature 30 °C, and carbon dioxide supply at 0.5 vvm. The light illumination was set at 2500 lux. The culture was harvested at end of exponential phase.

2.2 Extraction of Starch

Starch extraction was carried out using ethanol method [12]. Twenty five mg of microalgae biomass was extracted three times with 80% ethanol by vortex the samples for 1 minute then boil at 95°C for 10 minutes. After each extraction, the sample was centrifuged at 3500 rpm for 10 min. The sample was repeated extracted using ethanol until colorless samples to removes interfering substances such as pigments since it will affect the starch values. Then, the sample was wash twice with deionized water and acetone to remove ethanol residue. Then the cell was dried under fume hood for overnight before proceed for further analysis

.

2.3 Starch Analysis

The starch extracted then analyzed the purity using Megazymes total starch analysis kit. 10 mg of starch was used from the starch extracted. The total starch and the purity were calculated as follows:

$$Starch\ (\%) = \Delta A\ x\ \frac{F}{W}\ x\ FV\ x\ 0.9 \quad (1)$$

where, ΔA is absorbance reading against the reagent blank, F is 100 µg of D-glucose/absorbance for 100 µg of glucose (conversion from absorbance to µg), FV is final volume (10-100 ml) and W is weight (mg) of sample used.

$$\%\ Purity = \frac{m_{starch\ measured}}{m_{starch\ assayed}}\ x100 \quad (2)$$

2.4 Characterization of Starch

2.4.1 Moisture Content

The moisture content was determined using method described by Lee *et al.* [13] with modification by drying the starch sample at 110°C for 24 hours or until the weight became constant.

2.4.2Granule Morphology

The shape of starch granule was observed using scanning electron microscopy (SEM). The shape of granules was compared with corn starch granule [3].

2.4.3 Amylose Amylopectin Ratio

Amylose content was assayed using method used by Hassan *et al.* [14]. The starch sample of 0.1g was weighted into 100 ml of volumetric flask and added 1 ml 99% ethanol with 9 ml of 1M of sodium hydroxide (NaOH). The solution was mixed thoroughly then heated in boiling water for 10 minutes. After the solution was cooled, distilled water was added until the mark on volumetric flask and shaken to mix well. Then, 5 ml was pipette out from the sample and placed into new 100 ml of volumetric flask and 1ml of acetic acid and 2 ml of iodine solution were added to the new flask. Distilled water was added to the mark and shaken to mix well. The sample then was measured using spectrophotometer at absorbance 620 nm. The amylase and amylopectin content were calculated as below:

$$\text{Amylose content (\%)} = 3.06 \text{ x absorbance x } 20 \quad (3)$$

$$\text{Amylopectin (\%)} = 100 - \text{\% amylose content} \quad (4)$$

2.4.4Turbidity

The turbidity was determined using method used by Hassan *et al.* [14]. Starch suspension of 1% (w/v) was heated in water bath at 90°C for 1 hour with constant stirring. Then the sample was cooled for another 1 hour until reach 30°C. Then the sample was stored for 4 days at 4°C and the turbidity absorbance reading was determined for each 24 hour for 4 days using spectrophotometer at 640nm using water as blank.

2.4.5 Solubility and Swelling Power

Solubility (g/g, dry basis) and swelling power (%) of the starch was determined using method described by Nadiha *et al.* [15]. Starch suspension of 1% (w/v) was preheated in water bath at 50°C, 60°C, 70°C, 80°C, 85°C and 90°C for 1 hour then centrifuged at 2000 rpm for 20 min. The supernatant then poured into an eppendorf tubes then dried at 100°C in oven until it reach constant weight. Where the sediment was weighted after the supernatant poured out. The solubility and swelling power were calculated as follow:

$$SwellingPower\left(\frac{g}{g}\right) = \frac{Weight\ of\ the\ wet\ sediment\ (g)}{Weight\ of\ the\ dry\ starch\ (g)} \quad (5)$$

$$Solubility\ (\%) = \frac{Drysupernatant}{Drystarchmass} x\ 100 \quad (6)$$

3 Results

Klebsormidium flaccidum cell was harvested at the end of exponential as the highest starch accumulated at that phase. The biomass productivity of algae was 0.19 g/L.day. Total starch analyzed from the algae starch was 19.06% and commercial corn starch was 76.87%. The commercial corn starch was purchased as for market manufacturing for household. The starch powder produced based on optimized method for microalgae starch purification was characterized by 60% of purity. Microalgae species reported based on previous research, Rodjaroen *et al.* [16] was discovered for *Chlorella* sp. ranged of starch content 22% until 27%, *Chlorococcum* sp. was ranged 17% to 26%, *Scenedesmus* sp. reported the lowest from 7% until 23% starch content. The range was determined based on different species of the genus. Dragone *et al.* [17] had reported the starch content for *Chlorella Vulgaris* was 41%, the highest starch producer while Brányiková *et al.* [9] reported on *Chlorella vulgaris* able to produce almost 60% of starch content. Hence, in this study, *Klebsormidium flaccidum* was in the range of based on the previous reported microalgae starch and also could be extracted more the starch after undergo more purification process.

The moisture content of the samples shows the algae starch resulted in lower moisture content compare to corn starch as 24.8% and 50.8% relatively. Low moisture content makes easier storage as it could be stored in room temperature and less risk to bacterial contamination [18].

The shape of granule was observed under SEM in figure 1. The shape of starch from algae was observed to be relatively oval in figure 1(b) compare to corn starch shape was sphere showed in figure 1 (a). The aggregates of the algae starch was not made up from composite of other elements. The diameter of the granules were measured and resulted in the corn starch did had bigger granule size compare to algae starch. As shown in figure 1 (b) the diameter size of algae starch was measured nearly 1 µm which quite small as compare to granule from corn starch was nearly 7 µm. The application in industry for small granules now is growing interest in certain applications such as pharmaceutical, coating for fabric and film production [15]. The small granule is important in producing high quality of products due to its total specific area as the water uptake is increased and the hydration will lead to increase the viscosity, more sites for crosslinking reaction for starch modification and also high gelatinization temperature and resistance [3, 19]. The size of granules is influenced by the sources. The major sources of small granules could be found in sweet potato (5-10 µm), yam (5-15 µm), arrowroot (5-20 µm) that consists of 20% to 40% of small granules [20]. However, these sources has to undergo separation process which quite expensive and also since the sources are major source in food chain, hence limits the usage of these source [3]. The small granules found in algae would be the new source for the high quality of product.

IOP Conf. Series: Materials Science and Engineering **716** (2020) 012012 doi:10.1088/1757-899X/716/1/012012

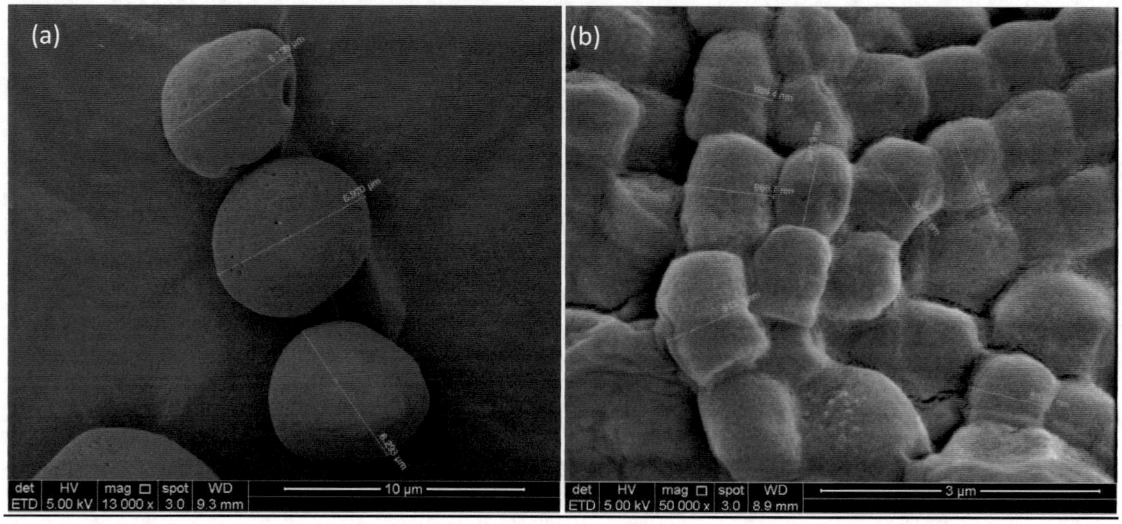

Figure 1. (a) Corn Starch granule was viewed under 13,000 magnifications with diameter of granules size and shape. 50,000 magnifications cannot be used on this sample since it will zoom too deep on granules. (b) *Klebsormidium flaccidum* starch granule was viewed under 50,000 magnifications with the diameter of granules and the shape. 13,000 magnifications cannot be used on this sample since it will show very small sizes of granules.

The amylose and amylopectin ratio was measured as the amylose content in algae starch was 25.52% and 74.48% of amylopectin content. Compare to corn starch which amylose content was 23.04% and the amylopectin was 76.96%. This shows almost similar content between corn and algae starch. In addition amylose/amylopectin ratio is mainly impacted by the biological origin and the environmental conditions which amylase content could be increased by controlling the microalgae culturing condition [3]. High amylase of starch content could lead to a strong and stiffer thermoplastic film with help by amylopectin that will form translucent paste that remains fluid when cooled [3, 14]. The turbidity of the corn starch was lower than the algae starch (figure 2). The turbidity measurement for each samples were increasing as the day of storage. This pattern was due to the swelling of granules and formation of network between amylase and amylopection chanins that were leached out of the granules during gelatinization [21].Turbidity is used to characterize the retrogradation behavior of diluted starch paste [22]. During the first 24 hours of storage, the rapid increase of turbidity was due to the amylose chains were leached out from the granules during gelatinization and reduce the transmission of light. Then after 48 hours, the turbidity was unchanged due to the aggregation of amylose and amylopectin chain was completed [22].

IOP Conf. Series: Materials Science and Engineering **716** (2020) 012012 doi:10.1088/1757-899X/716/1/012012

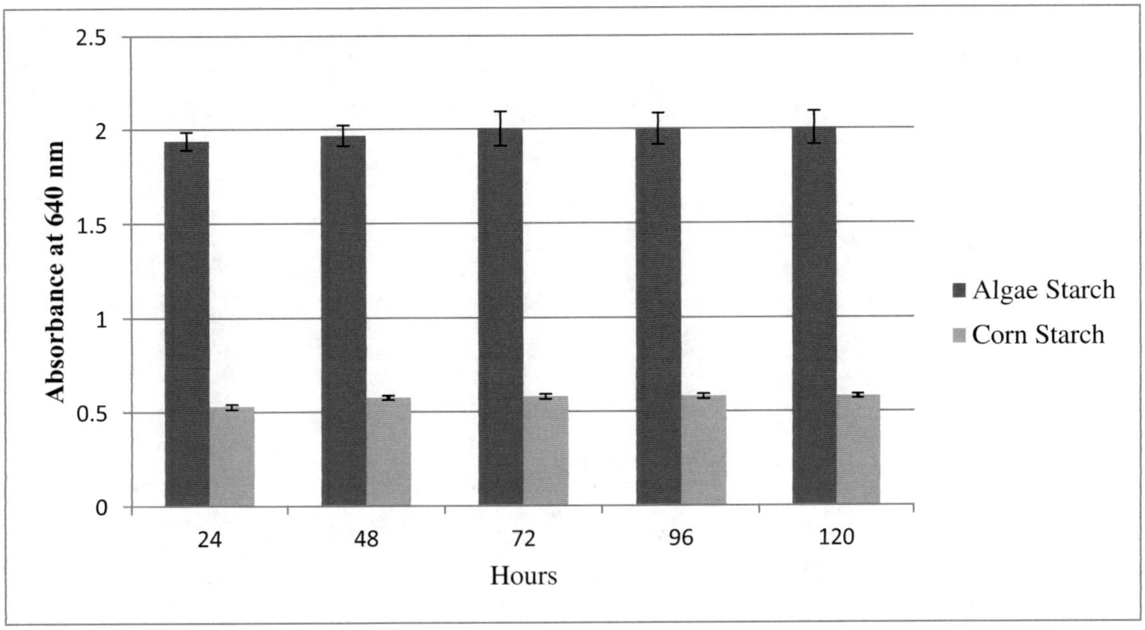

Figure 2. Turbidity of extracted starch from *Klebsormidium flaccidum* strain and commercial corn starch for 120 hours.

Swelling power is referred to the ability of water holding the starch granule and solubility is indicated as percent amount of starch leached out into the supernatant in the swelling volume determination [23]. The swelling capacity and water absorption happens when water adhered to the surface of starch granules and leads the granules to swell when exposed to higher temperature. The hydrogen bonds will be disrupted and replaced by water. The capacity of starch granules to swell will be depended on the capacity of starch molecules to hold water through the hydrogen bonding. This reaction is due to the amylose and amylopectin, given the water particles are trapped in amylopectin structure and also the difference in the distribution the chain length [24]. The granules size affects the gelatinization and the swelling capacity as the large granules will require more time to reach the gelatinization and small granules decreased it [25] due to the small granules have more efficient hydration than bigger one [26]. The hot water absorbs into the granules will caused them to swell. Then hydrogen bonds between water and hydroxyl groups of amylopectin and amylose are disrupted and trigger the swelling and solubilization of the starch granules. Subsequently, some of the amylose will be released from the granules and the granules will absorb more water and more swell [27].The swelling power of starch from algae was lower compare to corn starch [Figure 3(a)] and the solubility of the corn starch was higher than algae starch. This as could due to the amylose content and the granule size as described by the Cythia *et al.* [27]. The difference in swelling power is due to the different in amylose content, which biomass with lower amylose content will possess high swelling power and the solubility will depend on temperature as higher temperature will result in higher solubility.

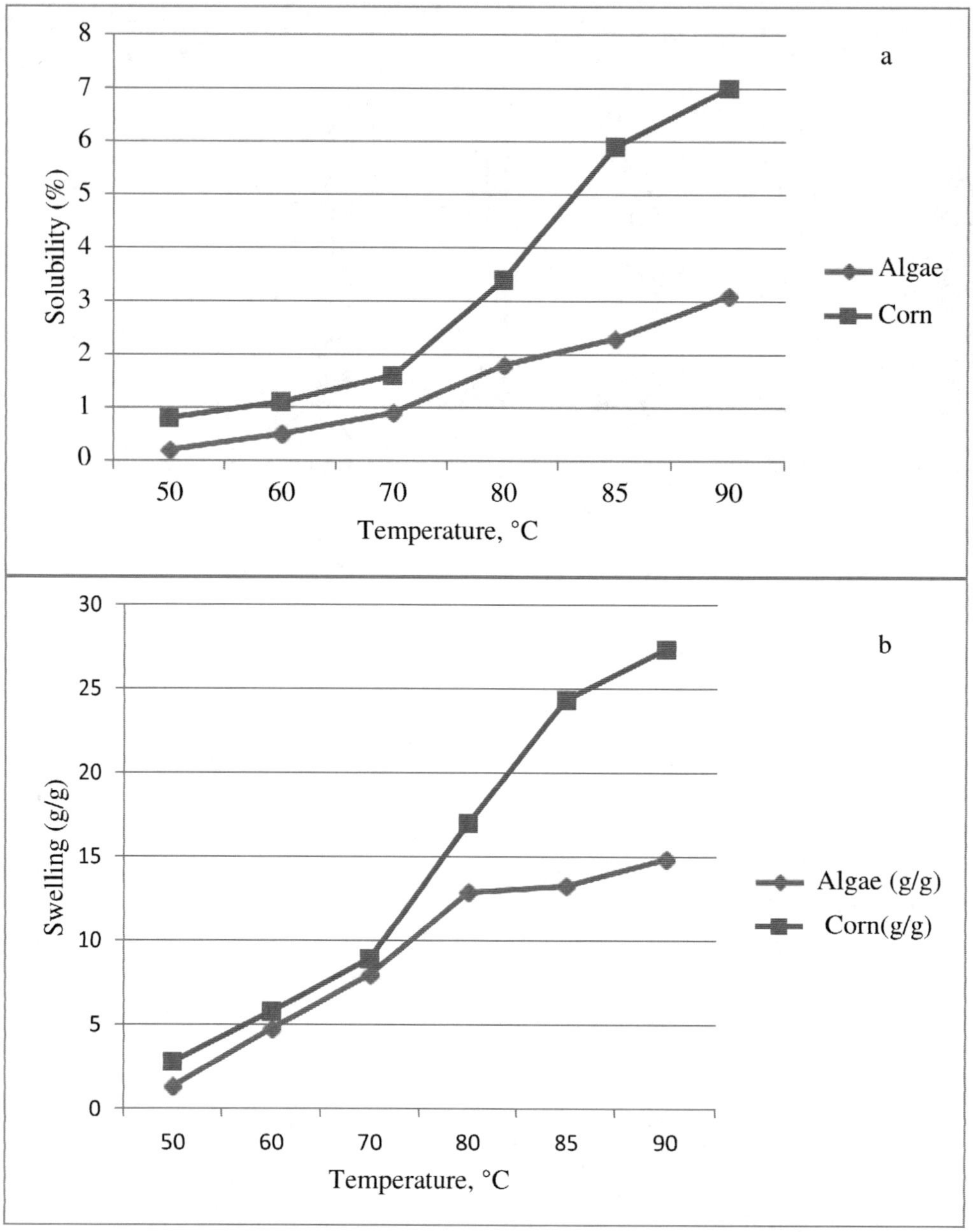

Figure 3. (a) Swelling power for the commercial corn starch and *Klebsormidium flaccidum*. (b) Solubility of the commercial corn starch and *Klebsormidium flaccidum*.

4 Conclusion

Microalgae as the new renewable source is a replacement for the current starch source. The advantages show by the characteristics studied as it showed interesting features for the industrial application such as in pharmaceutical, cosmetics. *Klebsormidium flaccidum* is a strain of marine microalgae of inexpensive, green sources that can be returned to its natural state without any pollution after product making. The features show in this study, the starch from *Klebsormidium flaccidum* strain is shows quite close characteristics with commercial corn starch and the structural properties are surpass the properties of corn starch. Therefore, more studies on the starch applications from *Klebsormidium flaccidum* strain would be a good evolution in industry.

References

[1] Sangwan, P., Petinakis, E., & Dean, K. (2014). Effects of Formulation, Structure, and Processing on Biodegradation of Starches. *Starch Polymers,* 357–378.

[2] Perezs., Bertoft E. (2010) The Molecular Structures of Starch Components and Their Contribution To The Architecture of Starch Granules: A Comprehensive Review. *Starch Journal*, **62**, 389-420.

[3] Gifuni I., Olivieri G., Russo Krauss I., D'Errico G., Pollio A., Marzocchella A. (2017) Microalgae as New Sources of Starch: Isolation and Characterization of Microalgae Starch Granules. *Chemical Engineering Transaction.***57**, 1423-1428.

[4] Harun, R., Danquah, M.K. and Forde, G.M. (2010) Microalgal biomass as a fermentation feedstock for bioethanol production. *J. Chem. Technol. Biotechnol.* **85**, 199–203.

[5] Giuliano D., Bruno D.F., Ana P.A., Antonio A. V., Jose A. T. (2011) Nutrient Limitation as a Strategy for Increasing Starch Accumulation in Microalgae. *J. Applied Energy*. 3331-3335.

[6] Sheehan, J., (2009) Engineering Direct Conversion of CO2to Biofuel. *Nat. Biotechnol.* **27**, 1128–1129.

[7] Shilton, A.N., Powell, N., Mara, D.D., Craggs, R., (2008) Solar-Powered Aeration and Disinfection, Anaerobic Co-Digestion, Biological CO2 Scrubbing and Biofuel Production: The Energy and Carbon Management Opportunities Of Waste Stabilization Ponds. *Water Sci. Technol.* **58**, 253–258.

[8] Laurens L.M., Dempster T.A., Jones H.D., Wolfrum E.J., Van Wychen S., McAllister J.S., Rencenberger M., Parchert K.J., Gloe L.M. (2012) Algal Biomass Constituent Analysis: Method Uncertainties And Investigation Of The Underlying Measuring Chemistries. *Anal Chem.* **84**:1879–1887.

[9] Brányiková I., Maršálková B., Doucha J., Brányik T., Bišová K., Zachleder V., Vítová M. (2011) Microalgae-Novel Highly Efficient Starch Producers. *Biotechnology and Bioengineering*, **108(4),** 766-776.

[10]A. Anderson, R. (2005) Algal Culturing Technique. London: *Elsevier Academic Press*, pp.117-130.

[11] Barsanti, L. and Gualtieri, P. (2014). *Algae: Anatomy, Biochemistry, and Biotechnology*. 2nd ed. Pisa, Italy: CRC Press Inc, pp.221-265.

[12] Chow, Pak & Landhäusser, Simon. (2004) A Method for Routine Measurements of Total Sugar and Starch Content In Woody Plant Tissues. *Tree physiology*. **24**. 29-36

[13] Lee, H. C.; Htoon, A. K. and Paterson, J. L. (2007) Alkaline Extraction of Starch from Australian Lentil Cultivars: Matilda and Digger Optimised for Starch Yield and Starch and Protein Quality. *Food Chemistry*, **102**, 551–559.

[14] Hassan L. G, Muhammad A. B., Aliyu R. U., Idris Z. M., Izuagie T., Umar K. J., Sanu N. A. (2013) Extraction and Characterisation of Starches from Four Varieties of *Mangifera Indica* Seeds. *Journal of Applied Chemistry.* **3(6)**, 16-23.

[15] Nadiha, M. Z. N., Fazilah, A., Bhat, R., Karim, K. A. (2010) Comparative Susceptibilities of Sago, Potato and Corn Starches to Alkali Treatment. *Food Chemistry,* **(121)**: 1053–1059.

[16] Rodjaroen, S., Juntawong, N., Mahakhant, A., & Miyamoto, K. (2007). High Biomass Production and Starch Accumulation in Native Green Algal Strains and Cyanobacterial Strains of Thailand *Kasetsart J. (Nat Sci),* **41**,570-575.

[17] Dragone, G., Fernandes, B. D., Abreu, A. P., Vicente, A. A., & Teixeira, J. A. (2011). Nutrient Limitation as A Strategy For Increasing Starch Accumulation In Microalgae. *Applied Energy,* **88(10),** 3331–3335.

[18] Ijabadeniyi, O. A., & Pillay, Y. (2017). Microbial Safety of Low Water Activity Foods: Study of Simulated and Durban Household Samples. *Journal of Food Quality,* **2017***,* 1–7.

[19] Lindeboom N., Chang P.R., Tyler R.T. (2004) Analytical, Biochemical and Physicochemical Aspects of Starch Granule Size, With Emphasis On Small Granule Starches.*A review, Starch Journal*, **56, (3-4),** 89-99.

[20] Peroni, F. H. G., Rocha, T. S., & Franco, C. M. L. (2006). Some Structural And Physicochemical Characteristics Of Tuber And Root Starches. *Food Science and Technology International,* **12(6),** 505–513.

[21] Ambigaipalan P, Hoover R, Donner E, Liu Q. (2013) Retrogradation Characteristics of Pulse Starches. *Food Res Int* **54**, 203 - 212.

[22] Wang, S., Li, C., Copeland, L., Niu, Q. and Wang, S. (2015) Starch Retrogradation: A Comprehensive Review. *Comprehensive Reviews In Food Science And Food Safety*, **14**, 568-585.

[23] Kaur, M., Oberoi, D. P. S., Sogi, D. S., & Gill, B. S. (2010). Physicochemical, Morphological and Pasting Properties Of Acid Treated Starches From Different Botanical Sources. *Journal of Food Science and Technology*, **48(4),** 460–465.

[24] Tang, H., Mitsunga, T., & Kawamura, Y. (2004). Relationship between Functionality and Structure in Barley Starches. *Carbohydrate Polymers,* **57**, 145–152.

[25] Koch, K., & Jane, J. (2000). Morphological Changes of Granules of Different Starches by Surface Gelatinization with Calcium Chloride. Cereal Chemistry, **77(2)**, 115–120.

[26] Moita, B. C., Lourenço, D. S. C. A., Bagulho, A. S., & Beirão-da-Costa, M. L. (2008). Effect of Wheat Puroindoline Allels on Functional Properties of Starch. *European Food Research Technology*, **226**, 1205–1212.

[27] Cynthia W., Quy D. D., Yi H. J., Shella P. S., Jindrayani N. P., Livy L., Felycia E. S., Suryadi I. (2019). Isolation and characterization of starch from Limnophila aromatic. *Heliyon*,**5(5),** 1-7.

Acknowledgements

The authors wish to thank School of Industrial Technology, Universiti Sains Malaysia and the Ministry of Higher Education, Malaysia (USM). The project and manuscript preparation were financially supported by the RUI Grant 1001/PTEKIND/811273.

IOP Conf. Series: Materials Science and Engineering **716** (2020) 012013 doi:10.1088/1757-899X/716/1/012013

Comparison of quantification methods and subsequent characterization of polyhydroxybutyrate film sample utilizing pretreated cane molasses as carbon source

Y S Khok[1], M Suwa[2], H Ito[2], M Hazwan Hussin[3], Y Ishida[2], K Sudesh[4], S Baidurah[1*]

[1]School of Industrial Technology, Universiti Sains Malaysia, 11800, Minden, Gelugor, Penang, Malaysia
[2]Department of Biological Chemistry, School of Bioscience and Biotechnology, Chubu University, 1200 Matsumoto-cho, Kasugai 487-8501, Japan

[3]School of Chemical Sciences, Universiti Sains Malaysia, 11800, Minden, Gelugor, Penang, Malaysia

[4]School of Biological Sciences, Universiti Sains Malaysia, 11800, Minden, Gelugor, Penang, Malaysia

*Email: sitibaidurah@usm.my

Abstract. Polyhydroxybutyrate (PHB) is a biodegradable polyester synthesized as intracellular carbon and energy storage for numerous bacteria and archaea. PHB has been used extensively in biomedical industry. The gas chromatography (GC) method for PHB quantification has some drawbacks, such as large sample size required and involves tedious sample preparation. In this study, reactive pyrolysis-gas chromatography (reactive Py-GC) was applied to determine the PHB content in bacterial biomass. *Bacillus* sp. was cultured in mineral salts medium containing various concentrations of pretreated molasses (10 – 30 g/L) as a sole carbon source. The PHB content obtained by both methods; GC and reactive Py-GC were compared. The obtained PHB was cast into film and the physical properties was determined by DSC and TGA. The biodegradability of PHB film sample was performed by soil burial biodegradation test. Based on the peak intensities of reactive Py-GC analysis, the PHB contents were determined precisely and rapidly without any cumbersome sample pretreatment. Furthermore, the values of the PHB contents coincided overall with those obtained by the conventional GC method with correlation coefficient (R^2) 0.9766. The melting temperature and degree of crystallinity were 172 °C and 61 %, respectively. The PHB film sample used in this study can biodegrade completely within 12 days.

IOP Conf. Series: Materials Science and Engineering **716** (2020) 012013 doi:10.1088/1757-899X/716/1/012013

1. Introduction

Polyhydroxybutyrate (PHB) is a type of biodegradable polyester thermoplastics that classified under polyhydroxyalkanoate (PHA) family. This biopolymer is synthesized by various types of bacteria, such as *Cupriavidus necator*, *Bacillus megaterium* and *Aeromonas caviae* under nutrient limiting conditions but in the presence of excess carbon source [1,2,3]. This biopolymer is accumulated in bacteria biomass in the form of cytosolic granule; serves as carbon and energy source for the bacteria during starvation [1,2,3]. The biopolymer can be extracted chemically or biologically from the bacterial biomass [1,3].

Over the past few decades, PHAs have gained many industrial attentions due to their biocompatibility, biodegradability and similar physical property to the commercial polypropylene [1]. Moreover, PHB is known to be degraded by the soil microorganism such as bacteria, archaea and fungi. However, the production cost for PHB is very high compared to petrochemical plastic [1]. Recently, many studies have been conducted to utilize agriculture by-product as the carbon source for the PHA production in order to improve its economic feasibility [2]. Cane molasses is one of the potential carbon sources for PHA production due to its abundance especially in Malaysia and high sugar content [2].

In terms of analysis method for PHB, gas chromatography (GC) is the standard method for PHB quantification; showing high accuracy and reproducibility. However, there are some drawbacks for GC method, for examples: (i) the use of hazardous solvent, (ii) tedious sample pretreatment, and (iii) large sample size required for the analysis. Therefore, it is required to develop a direct and highly sensitive method for PHB quantification [4].

On the other hand, reactive pyrolysis gas chromatography (reactive Py-GC) has been widely used in the forensic and archaeology field due to its rapid and direct method to analyse various polymeric materials, such as oil, polyaromatic compound and plastics [5,6]. The concept behind the reactive Py-GC is to use heat in controllable and reproducible way, and with the aid of strong organic alkali, such as tetramethylammonium hydroxide (TMAH) to breakdown high molecular weight sample selectively into smaller fragments before subjected to the pyrolyzer, attached at the inlet port of the GC for separation and detection. Based on previous studies, the authors have successfully applied the reactive Py-GC method to analyse the poly(3-hydroxybutyrate-*co*-3-hydroxyhexanoate) content in the bacterial cells directly, without any tedious sample preparation [6].

This study will compare the PHB content obtained by both conventional GC and reactive Py-GC analyses method. Then, the physical properties of PHB film samples produced by *Bacillus* sp. using pretreated molasses as main carbon source were characterized.

2. Materials and method

2.1. Material

The cane molasses used in this study was obtained from Malayan Sugar Manufacturing Company Berhad (MSM), Seberang Perai, Penang, Malaysia. The molasses solution was prepared by dissolving 15 g of molasses into 100 mL of distilled water.

2.2. Hydrothermal oven acid pretreatment

The hydrothermal oven acid pretreatment was performed by adding 10 mL of 1.5 N H_2SO_4 into 200 mL of molasses solution. The molasses solution was then transferred into 600 mL stainless steel hydrothermal reactor, preheat in the Venticell LSIS-B2V/VC55 hot air oven (München, Germany) at 105 °C for 1 hour and followed by incubation for another 1 hour. After the pretreatment, the pretreated molasses solution is cooled to room temperature and the pH were adjusted to 7.0. The pretreated molasses was centrifuge at 4500 rpm for 15 minutes to remove the impurities in the molasses before used for fermentation [2].

2.3. Microorganisms and media
The bacterial strains used in this study were *Bacillus* sp. The composition of the nutrient broth (NB) for inoculation (pH 7.0) was as follows (per L): 10 g meat extract, 10 g peptone, and 5 g sodium chloride. The mineral medium (MS) composition (pH 7.0) for *Bacillus* sp. was as follows (per L): 1 g$(NH_4)_2SO_4$, 1.5 g KH_2PO_4, 4.47 g $Na_2HPO_4 \cdot 2H_2O$, 0.2 g $MgSO_4 \cdot 7H_2O$ and 1 mL of trace element solution. The trace element solution was composed of (per L): 10 g $FeSO_4 \cdot 7H_2O$, 2.25 g $ZnSO_4 \cdot 7H_2O$, 1 g $CuSO_4 \cdot 5H_2O$, 0.5 g $MnSO_4 \cdot 5H_2O$, 2 g $CaCl_2 \cdot 2H_2O$, 0.23 g $Na_2B_4O_7 \cdot 10H_2O$, 0.1 g $(NH_4)_6Mo_7O_{24}$, and 10 mL 35% HCl. Pretreated molasses was used as main carbon source and test at various concentration ranging from 10 to 30 g/L [2].

2.4. Cultivation condition
Microbial culture was carried out at shake flask scale using one-stage cultivation strategy. The inoculation was done by cultivating the *Bacillus* sp. in nutrient broth (NB) medium at 30 °C, 200 rpm for 24 hours before transferring into the MM medium containing various concentration (10 – 30 g/L) of hydrothermal oven acid pretreated molasses. The cultivation conditions for *Bacillus* sp. are at 30 °C, 250 rpm for 2 days. After the fermentation, the cell biomass was harvested [2].

2.5 Extraction of polymer and film casting
The extraction of PHB was based on procedure reported by Murugan et al. [8], lyophilized cells were stirred in chloroform in a ratio of 1 g : 100 mL for 5 days at room temperature. The cell debris was filtered and the solution was concentrated using a Heidolph Laborata 4000 rotary evaporator (Schwabach, Germany) with the water bath set at 60°C. The concentrated solution was then added dropwise into 10 times volume of vigorously stirred cool methanol. The white precipitate, which is the PHB was recovered by filtration and air dried overnight. The solvent film cast method was performed according to Sridewi et al. [9]. The PHB film samples were prepared by dissolving the 0.3 g extracted polymer in 30 mL chloroform with magnetic stirring for 30 min. The mixture was cast into clean glass petri dishes (9 cm in diameter) and left to dry overnight.

2.6. Gas chromatography (GC) and reactive pyrolysis gas chromatography (reactive Py-GC) analysis for PHB determination
The PHB content in the cell biomass was analysed by GC and reactive Py-GC. For GC analysis, the sample was undergone methanolysis before subjected to GC. The sample preparation was conducted according to the procedure reported by Braunegg et al. [10]. Briefly, 2 mL of chloroform and 2 mL of methanol acidified with 15 % H_2SO_4 were added to the screw capped glass tube that contained 10 mg of lyophilized cell biomass. Then, the sample was heated on the dry bath at 100 °C for 3.5 hours. Upon the completion of methanolysis, the mixture was cooled to room temperature and 1 mL of distilled water was added to the mixture. The mixture was vortexed until 2 phases were formed, which are lower organic phase and an upper aqueous phase. The upper aqueous phase was removed, and sodium sulphate was added to remove any remaining water content. The amount of methyl 3-hydroxybutanoate was quantified using Shimadzu GC 2010 plus gas chromatograph (Kyoto, Japan) equipped with a flame ionization detector and a Merck SPB-1 capillary column (30 m length × 0.25 mm i.d., 0.25 μm film thickness). Helium was used as carrier gas with a flow rate 13.0 mL/min. The temperature program at 70 °C and ramp up to 280 °C at a rate of 5 °C/min [2]. In this procedure, standard sample was used to construct the standard calibration curve.

IOP Conf. Series: Materials Science and Engineering **716** (2020) 012013 doi:10.1088/1757-899X/716/1/012013

The sample preparation of reactive Py-GC was performed according to the procedure reported by Siti et al. [6,7]. The dry powder sample of bacteria colonies (0.1 mg) was weighed in the stainless steel sample cup (2 mm i.d. × 4 mm height) and 2 μL of 25 % tetramethyl ammonium hydroxide (TMAH) solution was added. The sample cup was dropped into the heated centre of Frontier Laboratory PY-2020iD vertical microfurnace pyrolyzer (Koriyama, Japan) that attached to the Shimadzu GC2010 Plus GC chromatography (Kyoto, Japan). The pyrolyzer was maintained at 400 °C under the carrier gas, helium with a flow rate 50 mL/min. A part of the flow (1 mL/min) reduced by the splitter was introduced into the Frontier Laboratory ULTRA ALLOY-5 (MS/HT) metal capillary separation column (30 m length × 0.25 mm i.d., 0.1 μm film thickness) coated with immobilised 5% diphenyl-95% dimethylpolysiloxane. The column temperature was programmed at 35 °C to 400 °C at a rate of 5 °C/min. The standard sample of PHB was used to construct the calibration curve [6,7].

2.7. Physical properties characterization of PHB film samples using differential scanning calorimetry (DSC) and thermalgravimetric analysis (TGA)

The differential scanning calorimetry analysis (DSC) was carried out using a Perkin Elmer DSC 6 instrument (Massachusetts, USA) under nitrogen flow of 20 mL/min. DSC analysis was conducted in triplicate which were heated from room temperature to 195 °C at a heating rate of 10 °C/min. The cooling cycle was performed from 195 to -10 °C. The degree of crystallinity for PHB was determined using Equation (1), considering the area of the melting peaks as the melting enthalpy:

$$X_c = \frac{\Delta H_m}{\Delta H_{0m}} \times 100\ \% \qquad (1)$$

where ΔH_m is the measured value for PHB melting enthalpy, ΔH_{0m} is the melting heat associated with pure crystalline PHB (146 J/mol) [11].

Thermal degradation measurement was performed using a Perkin Elmer Pyris 1 thermalgravimetric analysis (TGA) (Massachusetts, USA). Temperature was raised from room temperature up to 800 °C with a heating rate of 10 °C/min. Triplicate analyses were performed for PHB sample (5 mg). All runs were carried out under a nitrogen atmosphere (20 mL/min) in order to prevent any thermo-oxidative reaction.

2.8. Soil burial biodegradation test of PHB film sample

The PHB films were cut into pieces (1 cm × 1 cm) and cover with nylon cloth (3 cm × 3 cm) before buried in the compost soil with soil composition (58 % sand, 28 % silt and 14 % clay) including 13 % organic matter, at 30 °C, pH 7. Relative humidity was adjusted to approximately 70 % by using a soil moisture sensor, and water was added every 12 hours. All film samples were prepared in triplicate and buried in soil 3 cm from the surface until complete degradation was observed. The films were harvested at 2 days interval.

3 Results and discussion

Figure 1 shows the GC chromatogram of trans-methylated product of PHB after solvent extraction. After elution of the solvent peaks, the peaks of methyl 3-hydroxybutanoate which derived from trans-methylation of 3HB monomer was clearly observed on the chromatogram.

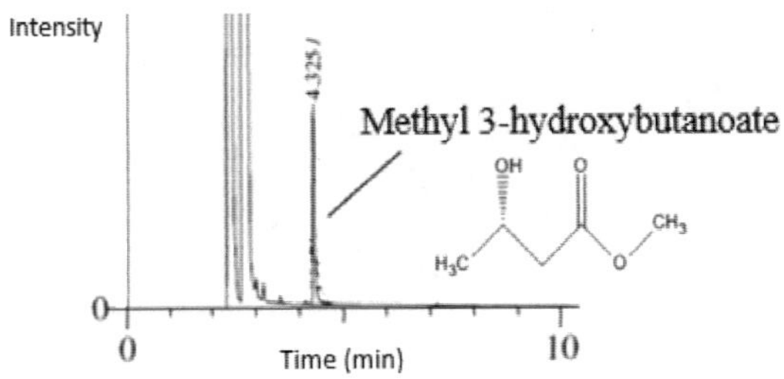

Figure 1. The typical GC chromatograms of *trans*-methylated product of PHB after solvent extraction.

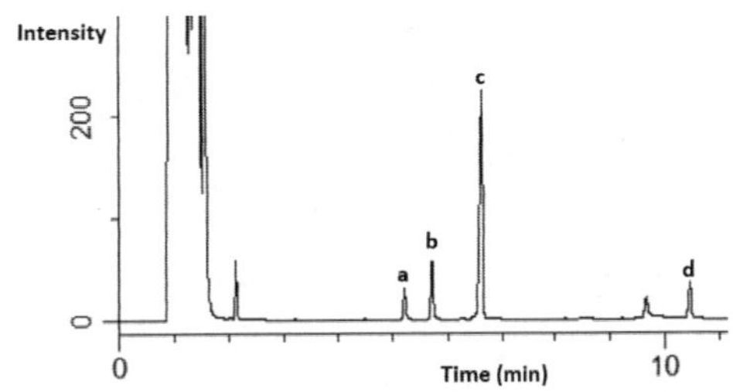

Figure 2 . The reactive pyrolysis gas chromatograms of PHB in the *Bacillus* sp.biomass without any solvent extraction.

Figure 2 shows the reactive pyrolysis gas chromatogram of PHB in the *Bacillus sp.* biomass without any solvent extraction. From the figure 2, four major peaks which attributed to methyl derivatives of 3HB monomer were clearly observed. The similar peak pattern was also reported by Siti et al., 2015 [7]. The assignment of these peaks were summarized in Table 1 along with their effective carbon number (ECN) corresponding to their molar sensitivity to the FID. Peak a-c had been identified as methyl 3-butenoate and its isomers; methyl *cis*-2-butenoate and methyl *trans*-2-butenoate while peak d had been identified as methyl 3-methoxybutanoate [6,7].

Table 1. Identification of the characteristic peaks on chromatogram figure 2

Peak number	Compound name	ECN[a]
a	Methyl 3- butenoate	3.65
b	Methyl *cis*-2-butenoate	3.65
c	Methyl *trans*-2-butenoate	3.65
d	Methyl 3-methoxybutanoate	3.95

[a] Effective Carbon Number for FID [12]

To calculate the PHB content in *Bacillus sp.* biomass, the following equation were applied, as proposed by Siti et al. [6,7] :

$$Y_{3\mathrm{HB}} = \sum_{i=1-3} \frac{\mathrm{I}_i}{\mathrm{ECN}_i} \tag{2}$$

where I_i is the area of peak *i* in Table 1, and ECN_i is the ECN corresponding to the peak *i*.

Figure 3 shows the comparison of PHB content accumulated in *Bacillus* sp. biomass using conventional GC and reactive Py-GC. The PHB content in *Bacillus sp.* biomass obtained from both methods show increasing trend that correlating to the increase of the concentration of pretreated molasses. To evaluate the accuracy of the reactive pyrolysis method, the PHB content of *Bacillus sp.* biomass obtained from GC and reactive Py-GC were compared. The PHB content obtained by reactive Py-GC is higher compare to GC method due to higher reaction efficiency of 3HB with TMAH, a similar result was also reported by Siti et al. [13]. This results indicate that the PHB content obtained by reactive Py-GC were strongly correlated with those by conventional GC. The correlation coefficient (R^2) between conventional GC and reactive Py-GC data was 0.9766, which indicate that the reactive Py-GC method can be serve as practical tool for direct and rapid PHB quantification in bacterial cell biomass.

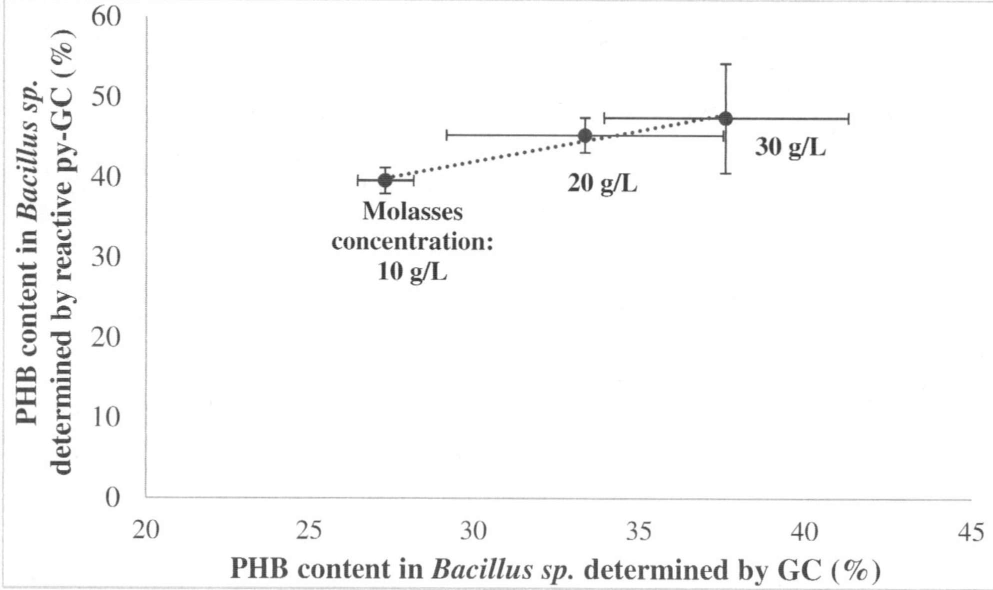

Figure 3. Comparison of PHB content accumulated in *Bacillus* sp. biomass using conventional GC and reactive Py-GC.

Figure 4 and Figure 5 shows the thermogram for both DSC and TGA for PHB films samples. The interpretation of physical properties for the PHB film samples are listed in Table 2. The glass transition temperature (T_g), which is -4.40 °C, is in the mentioned range in a previous study [13]. The T_g of PHB was depended on molecular factor, such as molar mass and degree of crystallinity (X_c), as well as the heating and cooling rate. Previous study reported that the T_g increase with the increase of molar mass of PHB up to 2×10^{-3} g/mol [15]. The crystallization temperature (T_c) obtained in this study is parallel as reported by Karami et al [18], which may due to the T_c was depend on the rate during cooling scan [19]. The melting temperature (T_m) of the PHB using pretreated molasses as the carbon source is 172.14 °C, which is consistent with the publications [16,17,20,22]. The decomposition temperature (T_d) of PHB film sample is 271.35 °C. These values are consistent with the results of PHB produced by *Rhodococcus equi* [17,20], which is mainly associated with the ester cleavage of PHB component by the β-elimination reaction [21]. The X_c and T_m of PHB in the present study are consistent with the results of PHB produced from *C. necator* cells when glucose is used as the carbon source [22]. Figure 4 shows the biodegradation of PHB film samples in term of weight reduction. From this figure, the PHB film samples can completely degrade within 12 days of soil burial biodegradation under the condition set in this experiment. The biodegradation of PHB in term of weight reduction was higher than in previous studies due to its degree of crystallinity and molecular weight of PHB, the soil types, organic content, including the bacteria and fungi consortium exists in the soil [23].

IOP Conf. Series: Materials Science and Engineering **716** (2020) 012013 doi:10.1088/1757-899X/716/1/012013

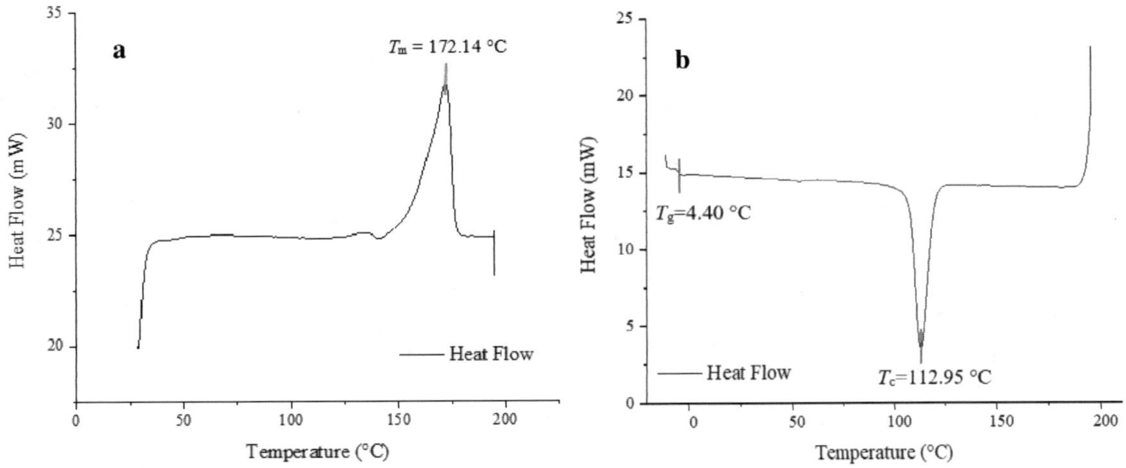

Figure 4. DSC thermograms of PHB film sample a) at the first heating, b) during the cooling scan.

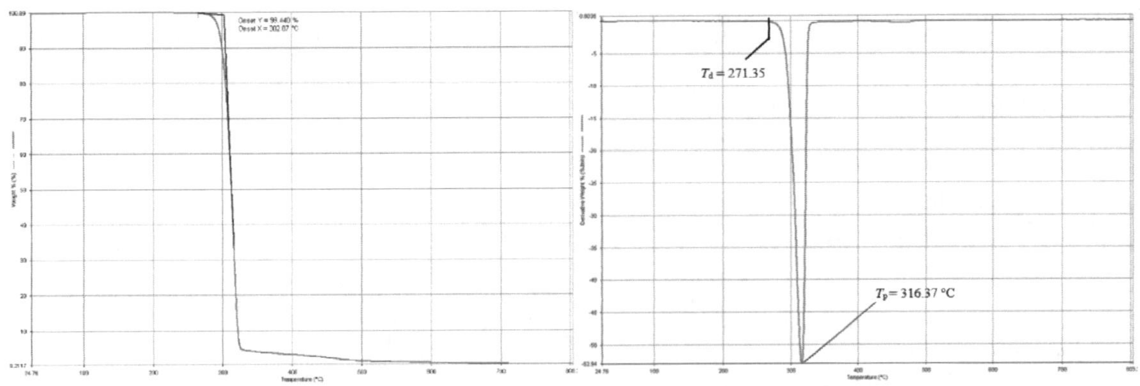

Figure 5. TGA thermogram of PHB film samples.

Table 2. Physical properties of PHB film sample

Physical Properties	This study	Altaee et al. [16]	Wang et al. [21]
Glass transition temperature, T_g (°C)	-2.42	2.79	-
Crystallization temperature, T_c (°C)	112.95	-	-
Melting temperature, T_m (°C)	172.14	173.00	172.05
Enthalpy of melting, ΔH_m (J/g)	90.50	-	89.70
Degree of crystallinity, X_c (%)	61.64	-	61.44
Degradation temperature T_d (°C)	271.35	276.00	249.60
Extrapolated onset temperature, T_o (°C)	302.87	-	-
Maximum weight loss temperature, T_p (°C)	316.37	-	269.80

IOP Conf. Series: Materials Science and Engineering **716** (2020) 012013 doi:10.1088/1757-899X/716/1/012013

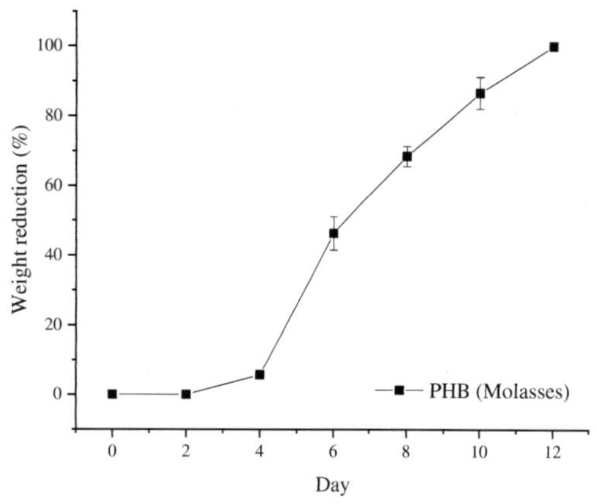

Figure 6. The biodegradation of PHB film sample in term of weight reduction.

4 Conclusion

Reactive Py-GC proves to be direct and highly sensitive method to determine the PHB content in the *Bacillus sp.* cell biomass without the need of tedious sample pretreatment and relatively small sample size (0.1 mg) required. Furthermore, the physical properties of PHB film samples produced by using the pretreated molasses as carbon source is similar to the previous studies.

5 References

[1] Sudesh K 2012 Polyhydroxyalkanoates from palm oil: biodegradable plastics (Heidelberg: Springer) p 1-32
[2] Khok Y S, Hussin M H and Baidurah S 2019 J Biocatal Agric Biotechnol **17** 51-59.
[3] Murugan P, Han L, Gan C Y, Maurer F H, and Sudesh K 2016 J Biotechnol **239** 98-105.
[4] Alves L P S, Almeida A T, Cruz L M, Pedrosa F O, De Souza E M, Chubatsu L S, Müller-Santos M and Valdameri G 2017 Braz J Med Biol Res **50** 1
[5] Wampler T 2014 Pyrolysis-Gas Chromatography *Reference Module In Chemistry, Molecular Sciences And Chemical Engineering* p 1-14
[6] Baidurah S, Murugan P, Joyyi L, Fukuda J, Yamada M, Sudesh K and Ishida Y 2016 *J Chromatogr* **1471** 186-191
[7] Baidurah S, Kubo Y, Kuno M, Kodera K, Ishida Y, Yamane T and Ohtani H 2015. *Anal Sci* **31** 79-83
[8] Murugan P, Gan C Y and Sudesh K 2017 *Int J Biol Macromol* **102** 1112-1119.
[9] Sridewi N, Bhubalan K and Sudesh K 2006 *Polym Degrade Stabil* **91** 2931–2940
[10] Braunegg G, Sonnleitner B and Lafferty R M 1978 *Eur J Appl Microbiol Biotechnol* **6** 29–37
[11] Seoane I, Manfredi L, Cyras V, Torre L, Fortunati E and Puglia D 2017 *Polymers*, **9** 561
[12] Jorgenson A D, Picel K C and Stamoudius V C 1990 *J Anal Chem* **62** 683–689
[13] Baidurah S, Kubo, Y, Ishida Y and Yamane T. 2018 *Pure Appl Chem*, **90** 1011-1017
[14] Srubar W V III, Wright C Z, Tsui A, and Frank C W 2012 *Polym Degrad Stab* **97** 1922–1929
[15] Androsch R, Radusch H J and Funari S S 2007 *Eur Polym J* **43** 4961–4974
[16] Pachekoski W M, Agnelli J A M and Belem L P 2009 *Mater Res* **12** 159-164.

[17] Altaee N, Fahdil A, Yousif E and Sudesh K 2016 *J Taibah Univ Sci* 10 543-550.
[18] Karami S, Ahmadi Z, Nazockdast H, Rabolt J F, Noda I and Chase B D 2018 *Mater Res Express* **5** 015316.
[19] Wellen R M R, Rabello M S, Araujo Júnior I C, Fechine G J M and Canedo E L 2015 *Polímeros* **25** 296-304.
[20] Zhang M and Thomas N L 2011 *Adv Polym Tech* **30** 67-79
[21] Choi J Y, Lee J K, You Y and Park W H 2003 *Fiber Polym* **4** 195-198.
[22] Wang Y, Chen R, Cai J, Liu Z, Zheng Y, Wang H, Li Q and He N 2013 *PLoS One* **8** e60318.
[23] Kijchavengkul T and Auras R 2008 *Polym Int* **57** 793–804

Acknowledgement
The authors would like to express their gratitude toward MSM Malaysia Holding Berhad, Kwok Foundation and Universiti Sains Malaysia RUI Grant 1001/PTEKIND/8011022 for financial support of this study.

IOP Conf. Series: Materials Science and Engineering **716** (2020) 012014 doi:10.1088/1757-899X/716/1/012014

Sugars from sago frond as prebiotic substrate to enhance the growth of *Lactococcus lactis* IO-1 and production of L-lactic acid.

M N Ahmad[1], K B Bujang[1], D S A Adeni[1] and F Lananan[2]

[1]Biotechnology Programme, Faculty of Resource Science and Technology University Malaysia Sarawak (UNIMAS) 94300, Kota Samarahan, Sarawak, Malaysia
[2]School of Biotechnology Universiti Sultan Zainal Abidin (UniSZA) Kuala Terengganu, Terengganu, Malaysia.

Email: nrwhelmi92@gmail.com

Abstract

Sago palm is often discredited for exhibiting long maturity period and barrenness of pre-harvest products, which restrain its potentials as an alternative and eternal starch provider. The use of sago fronds to produce prebiotic and fermentable sugars from pruned palms and fronds discarded upon harvesting is a possible enterprise to provide income for the cash-strapped sago farmers while waiting for the sago trunks to be harvestable. Dried sago frond powder coupled with the cellulolytic enzyme and incubated for 48 hours, producing a maximum recovery of cellobiose at 25%. This is of great advantage in reducing the cost of large-scale processes since the yield and productivity from SFS is comparable to the Standard Medium and SFS amended with yeast extract at 0.85g/g and 85%, respectively. Meanwhile, the composition of cellobiose as main sugar component increase the viability of the *Lactococcus lactis* I0-1 by prolong the lifespan of the cell by perform as slow release carbon source, in fact, cellobiose was protected by β (1-4) glycosidic bond made it consumable to specific probiotic in human digestive system conceive that cellobiose as potential prebiotic component for human. Clearly, the use of sago frond is highly economical and sustainable as the raw material for the manufacturing of fermentable sugars and subsequently as the sustainable substrate for large-scale production of L-lactic acid.

IOP Conf. Series: Materials Science and Engineering **716** (2020) 012014 doi:10.1088/1757-899X/716/1/012014

1.0 INTRODUCTION

Lactic acid is widely used in the food industry (as a preservative), in the cosmetic industries (skin and hair care products), and in the pharmaceutical and textile industries. Lactic acid can be produced both by microbial fermentation and synthetically. Microbial fermentation using lactic acid bacteria and biological wastes is highly regarded since it utilizes renewable resources and does not emit greenhouse gasses especially carbon dioxide [8]. *Lactococcus lactis* IO-1 is the homolactic bacteria widely used for the production of lactic acid due to ability of the strain produce L-(+) lactic acid as the main product [9]. Generally, lactic acid is produced using starch from food crops such as maize, sugar cane and tapioca as the main raw material. However, production of lactic acid from food crops for use other than as food preservatives is deemed unethical and hence must be avoided.

Consequently, alternative raw materials for the production of lactic acid such as agricultural wastes should be encouraged to ensure sustainability which guarantees production and concomitantly minimise pollution. Therefore, in this research the use of sago frond as the alternative raw material for the production of lactic acid is proposed due to the humongous amount of this waste generated upon harvesting of sago trunks for the production of sago flour. Utilization of sago frond for the production of lactic acid not only will reduce environmental problems due to the slow degradability of sago fronds but will also increase the sago farmers' income, while denying ethical issues of producing lactic acid from food crop commodity.

2.0 MATERIALS AND METHODS

2.1 Materials

2.1.1 Sago Frond Sugar (SFS)

Sago fronds were obtained from locally cultivated sago palms in the District of Pusa, Sarawak. Sago Frond Sugar was produced through enzymatic hydrolysis of sago frond. Deskined sago frond was cut into small 2-3 cm cubes, oven dried and ground to produce powder, name as Sago Frond Powder (SFP). About 10% (v/w) of the enzyme Celluclast 1.5L (60 FPU/mL) was used to convert cellulose in SFP into SFS. The hydrolysis was conducted at 45°C for 48 hours. Detail procedure where treated sago frond fibre has been enzymatically hydrolysed using the cellulolytic enzyme producing Sago Frond Sugar (SFS), which mainly contains cellobiose and glucose at 15-17 g/L and 5-7 g/L, respectively [2].

2.1.2 Microorganism and Standard Media

The strain of *Lactococcus lactis* IO-1 was obtained from the Japanese Collection of Microorganism (JCM) with an optimum growth temperature at 37°C [10]. Stock culture was revived in Nutrient Broth (NB) media for 18 hours at 37°C. Exactly 10% (v/v) of the culture was transferred to the inoculum medium.

2.1.3 Inoculum

A 10% (v/v) inoculum was used as standard in this experiment. The inoculum medium consisted of 20 g/L commercial glucose and 5 g/L yeast extract. Once the medium was sterilized, the inoculum was incubated for 18 hours at 37°C with agitation. The incubated inoculum was divided into 3 sterile falcon tubes at 10% (v/v) of working volume, and centrifuged (8,000 rpm) at 4°C. The supernatant from the inoculum was discarded.

***2.2** Fermentation*

The experiment was conducted using batch fermentation system in 2L shake flasks at 1L working volume. Three types of fermentation media were used in this experiment, these are Standard Medium (SM) made of commercial glucose and yeast extract at 20 g/L and 5 g/L respectively, SFS supplemented

IOP Conf. Series: Materials Science and Engineering **716** (2020) 012014 doi:10.1088/1757-899X/716/1/012014

with yeast extract (SFS+YE) and SFS, without the addition of yeast extract. The fermentation media were incubated for 48 hours at 37°C with the agitation speed of 150 rpm. Exactly 10ml samples were collected into pre-weighed centrifuge tubes for every 6 hours to determine the cell growth profile, sugar consumption and production of lactic acid.

2.3 Growth Profile

The growth profile of *L. lactis* IO-1 was determined using the dry cell weight (DCW) method. Samples collected from the fermentation process were centrifuged (10,000 rpm) at 4°C. The supernatant was placed into new falcon tubes to determine the sugar consumption and lactic acid production. The cell pellets were dried in oven at 60°C until the weight is constant.

2.4 Sugar Consumption and Lactic Acid Production

The supernatant obtained from the fermentation process was analysed for sugar and L-lactic acid using High Performance Liquid Chromatography (Shimadzu, Japan) with BioRad Fermentation Monitoring Column at 1.0 ml/min flowrate using 0.005M H_2SO_4 as mobile phase at 65°C column temperature.

2.5 Statistical Analysis

All data obtained from all of experiments were compared and analysed using a One Way and Two-Way ANOVA. LEAD Technologies, MINITAB Version 17.1.0 was used to perform the statistical analyses and all graphs were generated using OriginPRO Version 9.0.0 software by OriginLab.

3.0 RESULTS AND DISCUSSION

3.1 Growth Profile of L. lactis IO-1 in SM, SFS+YE and SFS

The growth of *Lactococcus lactis* IO-1 in SM showed that the lag phase commenced in less than 6 hours upon incubation, followed by the exponential phase which remains up to 12 hours later (figure 1). The growth reached stationary phase after 18 hours with a biomass (DCW) of 1.60 g/L.

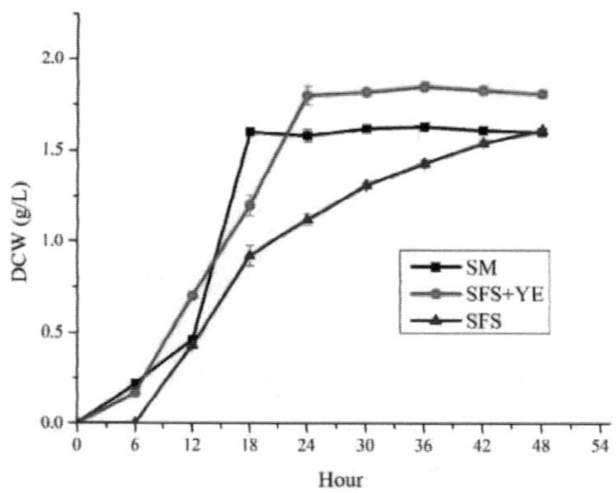

Figure 1: Growth profile of *L. lactis* IO-1 in different types of media; standard medium (SM), sago frond sugar amended with yeast extract (SFS+YE) and only sago frond sugar (SFS), based on dried cell weight (DCW, g/L).

In the medium where SFS was amended with yeast extract (SFS+YE), the growth of *Lactococcus lactis* IO-1 showed the lag phase started in less than 6 hours, similar to the SM, following which the cell grew exponentially for the remaining 18 hours in the lag phase. Stationary phase was reached after 24 hours of incubation with a biomass (DCW) of 1.80 g/L.

Comparatively, in the medium where SFS was used as the sole substrate (SFS medium), *Lactococcus lactis* IO-1 remained in the lag phase for 6 hours. But after this, the biomass promptly increased and remains in the lag phase for 12 hours of exponential growth. Based on this observation, the cells cultured in SFS continued to proliferate even after 48 hours of incubation, where the amount of biomass (DCW) was 1.61 g/L. Evidently, incubation of *Lactococcus lactis* IO-1 in SFS+YE produced the highest DCW (1.80 g/L) compared with SM and SFS, which was both at about 1.60 g/L. Utilization of cellobiose as the main carbon source enhanced the growth of *Lactococcus lactis* IO-1 but it needs longer incubation time.

Analyses of the kinetic parameters of the bacterium cultured in SM, SFS+YE and SFS (table 1) shows that the productivity of *L. lactis* IO-1 in SFS+YE medium (0.04^{b}) was only slightly lower compared to SM (0.05^{a}). This also confirmed that cellobiose in SFS requires a little more time for the bacterium to be used as substrate.

Table 1: Kinetic parameters of *Lactococcus lactis* IO-1 grown in SFS+YE and SFS media.

	Types of Medium		
Parameters	SM	SFS+YE	SFS
Maximum DCW Concentration (g/L)	1.6^{b}	1.8^{a}	1.61^{b}
Specific Growth Rate/ μ (h^{-1})	0.03^{b}	0.04^{a}	0.03^{b}
Yield (g/g)	0.07^{b}	0.08^{a}	0.08^{a}
Productivity (g/L/hour)	0.05^{a}	0.04^{b}	0.03^{c}

Statistical analysis was determined using One-way ANOVA and Tukey tests. Each value is the mean ± SE of 3 replicates. Means with the same letter are not significant at $p < 0.05$

Lactococcus lactis ssp. has a phosphoenolpyruvate dependent transport system (PTS) for fructose, mannose, sucrose or trehalose, mannitol and cellobiose [4]. This characteristic consents *L. lactis* IO-1 to utilize cellobiose as the carbon source. However, excessive cellobiose extremely inhibits glucose uptake up to 81% [11]. Hence, *L. lactis* IO-1 depend mainly on cellobiose as the carbon source. In order to utilize cellobiose, *L. lactis* IO-1 needs to hydrolyse cellobiose into glucose before it can be consumed for energy. Therefore, the productivity of *L. lactis* IO-1 in medium supplemented with cellobiose as the carbon source is slightly lower with longer lag phase compared to growth in medium using glucose as the carbon source. Consequently, cellobiose supplementation enhances the growth of *L. lactis* IO-1 as proven in prebiotic analysis. The specific growth rate (0.04 h^{-1}) and the yield (0.08 g/g) of cell biomass of *L. lactis* IO-1 in SFS+YE media slightly higher compared to SM (0.03 h^{-1} specific growth rate, 0.07 g/g yield). Slow degradation of cellobiose as carbon source by *L. lactis* IO-1 allowed the stain to undergo longer exponential period and multiply more at higher growth rate lead to higher yield of biomass.

Based on these results, the lag phase in SFS+YE medium was shorter compared to SFS medium. This shows that supplementation with YE have boosted the growth of *L. Lactis* IO-1, similar to the finding of [15]. Likewise, the lag phase of *L. lactis* IO-1 in SFS medium was much longer compared to

SFS+YE medium. This is because of the productivity and growth rate of cells in SFS+YE are higher compared to in SM. Higher productivity and growth rate mean increased growth of *L. lactis* IO-1. Hence the amount of biomass of *L. lactis* IO-1 in SFS+YE was higher compared to SFS.

Supplementation of yeast extract was important for fast growth of *L. lactis* IO-1 and subsequently the concentration of biomass, potentially reducing the duration of fermentation [15]. In the absence of yeast extract, the growth of *L. lactis* IO-1 in SFS media was comparable to SM. There was no significant different in maximum DCW and growth rate of *L. lactis* IO-1 in these two media, in fact the yield of biomass in SFS was higher compared to SM. This mirrors the potential of cultivation *L. lactis* IO-1 in SFS without the amendment with yeast extract.

Sago frond contains natural nitrogen source, at approximately 2.26% DM. In our project, 6% w/v of dry matter was used for the enzymatic hydrolysis process (60 g sago frond powder in 1L mixture), containing about 1.35 g of nitrogen which naturally exists in the fermentation medium [6]. Hence, SFS have high potentials to be used as an inoculum or a fermentation medium to grow *L. lactis* IO-1.

Aerobic fermentation of *Lactococcus lactis spp.* require superoxide dismutase (SOD) to utilize oxygen free radical (O^-_2) into hydrogen peroxide (H_2O_2) by via dismutation process [17]. The present of hydrogen peroxide as free radical compound inhibits the growth of *L. lactis.* Hence, catalase enzyme will be produced to convert hydrogen peroxide into water and stable oxygen (O_2) molecule.

Some of the nutrients are important in enzymatic reaction to enhance the activity of the enzyme that responsible as cofactor mineral. Superoxide dismutase form can be influence by the present of zinc, copper and manganese [14]. These minerals found in raw SFS which can stimulate the activity of SOD produced by *L. lactis* IO-1.

Lactococcus lactis require at least 0.06 mg/L of manganese in the fermentation media to stimulate the activity of superoxide dismutase enzyme in aerobic fermentation [7]. For addition, the concentration of magnesium lower than 0.36mg/L affect the growth rate of the *L. lactis,* nevertheless the final biomass produced were reliant on the initial Mg concentration of the fermentation media that contain lower than 2.4 mg/L [12]. Both nutrients (Appendix C) exits naturally in SFS and the amount was adequate (Mn = 18.37 mg/L, Mg = 72.66 mg/L) for the growth of *L. lactis* IO-1 in the production of lactic acid.

Due to the presence of these nutrients, production cost of lactic acid via fermentation of SFS using *L. lactis* IO-1 can be reduced due to unnecessarily supplementation of nitrogen sources such as yeast extract which cost up to RM 4,775/tonne for fermentation of SFS.

SOD is a very important antioxidant in neutralizing thousands of toxic free radical and for the production of anti-radical medicine and prevent free radical skin damage in cosmetic products [16]. Therefore, SOD produced from the fermentation of *L. lactis* IO-1 in SFS can be extracted and purified or combined with lactic acid for the production of high-value product which will enhance development of the sago industry.

3.2 Consumption of Sugar and Production of Lactic Acid

It was observed that maximum concentration of lactic acid produced from the fermentation of *L. lactis* IO-1 in SM was 22.81 g/L±0.06 after 42 hours (figure 2). The amount of residual sugar was 0.85 g/L±0.02. Hence, the amount of sugar consumed was 22.51 g (93.79%) with a lactic acid yield of 94.21%. The highest production of lactic acid in SFS+YE medium was 21.40 g/L (89.16% yield) with a residual sugar of 0.91 g/L (96.21% sugar consumed) after 48 hours. Within the same duration (48 hours of incubation) the highest amount of lactic acid in SFS medium was 18.64 g/L±0.06 (80.8% yield) and final sugar concentration of 4.23 g/L±0.23 (81.66% sugar consumed).

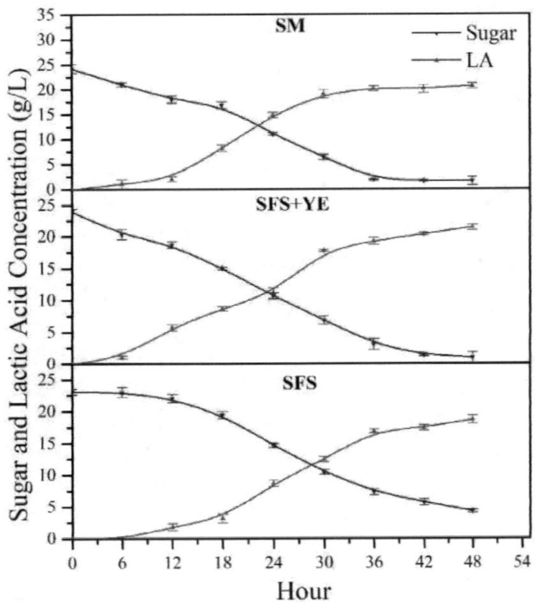

Figure 2: Comparison in production of Lactic acid by *L. lactis* IO-1 in different media.

Evidently in this project, all parameters are superior in SM medium compared to the other two media. The yield of lactic acid from SM medium was higher compared to SFS+YE (table 2), since the fermentation of glucose has higher conversion rate to lactic acid than cellobiose (as in SFS+YE and SFS). The efficiency of lactic acid production in SM was higher compared to SFS+YE and SFS media together with shorter duration of fermentation (36 hours in SM), faster by 12 hours than SFS+YE (48 hours).

Table 2: Kinetic parameter of L-lactic acid production of *L. lactis* IO-1 in SM, SFS+YE and SFS media.

	Types of Medium		
Parameters	SM	SFS+YE	SFS
Yield (g/g)	0.91^{a}±0.008	0.85^{b}±0.006	0.85^{b}±0.001
Productivity (g/L/hour)	0.64^{a}±0.004	0.45^{b}±0.009	0.39^{c}±0.001
Efficiency (%)	92.40^{a}±0.77	85.03^{b}±0.63	84.76^{b}±0.85

All kinetic parameters showed that glucose is the better substrate for the production of lactic acid. This is because of the molecular structure of cellobiose is more complex than glucose. Glucose can be directly utilized into lactic acid by lactic acid bacteria such as *L. lactis* IO-1 while cellobiose was protected by the β (1-4) glycosidic bond that must be cleaved into glucose molecules using the enzyme β-glucosidase before it can be converted to lactic acid. This needs more time and reduced the convertibility of cellobiose into lactic acid by *L. lactis* IO-1. *L. lactis* strains that encode pBL1 and celB genes produce a phospho-β-glucosidase for the hydrolysis of cellobiose indicate the growth profile of strain grown in chemically defined medium containing cellobiose (CDM-Cel) is slower compared glucose as carbon source due to existence of those gene that shifted the fermentation products toward a mixed acid profile and promoted significant changes in intracellular pool size for glycolytic intermediates in cells growing on cellobiose consumption [5]. Moreover, cultivation of *L. lactis* strain in CDM-Cel fro 24 hours shows a very long lag phase that resulted in the increase the culture optical density [13].

However, utilization of SFS as the alternative medium for the production of lactic acid is very much akin to SM. It is possible to have incomplete conversion of cellobiose to glucose via enzymatic hydrolysis, but addition of more β-glucosidase enzyme was unnecessary due to convertibility of cellobiose to lactic acid by *L. lactis* IO-1. The results obtained was similar to [1] where the lactic acid recovery from the utilization of pure cellobiose was 90% using *Lactobacillus delbrueckii* mutant Uc-3 strain after 40 hours of incubation. This shows that utilization of cellobiose as the main and cheaper source of carbon in SFS is capable for large-scale production of lactic acid.

The productivity of lactic acid from SFS+YE medium was higher compared to SFS, which supports that addition of yeast extract speed up the production of lactic acid by enhance the growth of *L. lactis* IO-1. However, rapid growth of *L. lactis* IO-1 in SFS+YE medium creates a stress environment to the cells due to increased biomass concentration which leads to competition for the carbon source. This competition forced the bacteria to produce lactic acid to exterminate other bacteria [3]. As such, productivity of lactic acid in SFS+YE medium is higher compared to SFS. Low productivity will lead to slower production of lactic acid. Hence, production of lactic acid and consumption of sugar in SFS medium was incomplete after 48 hours of incubation.

However, the yield and efficiency of lactic acid in SFS and SFS+YE medium was not significantly different. This proves that addition of yeast extract does not influence the conversion of cellobiose to lactic acid and it does not influence the conversion of cellobiose to glucose. This is because nitrogen source naturally exist in SFS. Thus, supplementation of yeast extract only speeds up the conversion of sugar to lactic acid. Therefore, production of lactic acid from the utilization of SFS can be conducted efficiently without the need to add yeast extract, it only requires slightly longer incubation time.

4.0 CONCLUSION

In conclusion, sago frond has high potential to be utilized as alternative raw material as viable prebiotic media to enhance growth of lactic acid bacteria (LAB) and excellent substrate for production of lactic acid. Large-scale production of lactic acid is then possible, without the need to interfere with production of starch food products and crops such as maize, sugar cane and sago starch. Lactic acid produced from the fermentation of SFS by using *L. lactis* IO-1 strain is purely of the L-lactate type, which has higher values for use as food preservatives and cosmetics.

REFERENCES

[1] Adsul, M., Khire, J., Bastawde K. and Gokhale, D. (2007). Production of Lactic Acid from Cellobiose and Cellotriose by *Lactobacillus delbrueckii* Mutant Uc-3. *American Society for Microbiology*, **73** (15)

[2] Adeni, D. S. A., Ahmad, M. N., & Bujang, K. (2018). Maximising production of prebiotic sugar (Cellobiose) from sago frond. *Malaysian Applied Biology*, ***47***, 89–96

[3] Bujang, K. B., D.S.A and Ang, S.Y. (2005). Production and Purification of L-Lactic Acid from Fermentation of Hydrolysed Sago Starch. Proceeding of the 27th Malaysian Microbiology Symposium. Penang. 24th- 27th Nov., 2005

[4] Bolotin, A., Wincker, P., Mauger, S., Jaillon, O., Malarme, K., Weissenbach, J. and Sorokin, A. (2001). The Complete Genome Sequence of the Lactic Acid Bacterium. *Genome Research*, 731–753.

[5] Campelo, A. B., Gaspar, P., Roces, C., Rodríguez, A., Kok, J., Kuipers, O. P., … Martínez, B. (2011). The Lcn972 Bacteriocin-Encoding Plasmid pBL1 Impairs Cellobiose Metabolism in *Lactococcus lactis* , ***77*(21)**

[6] Flach, M. and Schuiling, D.L. (1991). Growth and Yield of Sago Palms in Relation to Their Nutritional Needs. Proceeding of Fourth International Sago Symposium, Kuching, Sarawak, Malaysia. Pp 103-110.

[7] Hansson L, Haggstrom MH (1984) Effects of growth conditions on the activities of superoxide dismutase and NADH-oxidase/ NADH-peroxidase in *Streptococcus lactis*. *Current Microbiology,* **10,** 345-352**.**

[8] Ishizaki, A. (1997). Concluding Remarks for the Sixth International Sago Symposium at Riau. S*ago Communication*, **8,** 22-24.

[9] Ishizaki, A. and Okta, T. (1989). Batch Culture Kinetics of L-lactate fermentation employing *Streptococcus sp.* IO-1. *Journal of Fermentation and Bioengineering*, **67**, 46-51.

[10] Ishizaki, A., Osajima, K., Nakamura, K., Kimura, K., Hara, T. and Ezaki, T. (1990). Biochemical characterization of *Lactococcus lactis* IO-1 whose optimal temperature is as high as 37C. *J. Gen. Appl. Microbiol.* **36:** 1-6.

[11] Kim, O. B., Richter, H., ZaunmüLler, T., Graf, S. and Unden, G. (2011). Role of secondary transporters and phosphotransferase systems in glucose transport by Oenococcus oeni. *Journal of Bacteriology*, ***193***(24), 6902–6911.

[12] Loubiere P., Cocaign-Bousquet M., Matos J., Goma G. and Lindley N.D. (1997). Influence of end-products inhibition and nutrient limitations on the growth of *Lactococcus lactis* subsp. lactis. J Applied Microbiology **82**, 95-100.

[13] Solopova, A., Kok, J., & Kuipers, O. P. (2017). Disruption of a Transcriptional Repressor by an Insertion Sequence Element Integration Leads to Activation of a Novel Silent Cellobiose Transporter in *Lactococcus lactis* MG1363. *Applied and Environmental Microbiology*, ***83***(23)

[14] Sorochan, W. (2016). Supercharging the Immunity System. Retrieved from http://www.freegrab.net/glisodin.htm.

[15] Tanaka, K., Ohta, T. and Ishizaki, A. (1995). Effect of nitrogen sources in culture medium on L-lactate fermentation employing *Lactococcus lactis* IO-1. *Japan Faculty of Agriculture, Kyushu University*, ***39***, 131–138.

[16] Wang, J., Liu, J., Lu, J., & Cui, Z. (2010). Isolation and purification of superoxide dismutase from garlic using two-stage ultrafiltration. *Journal of Membrane Science*, ***352***(1–2)

[17] Van Niel, E. W. J. and Hahn-Hägerdal, B. (1999). Nutrient requirements of lactococci in defined growth media. *Applied Microbiology and Biotechnology*, ***52***(5), 617–627.

ACKNOWLEDGEMENT

This research has been funded by the Fundamental Research Grant Scheme (F07/FRGS/1613/2017) from the Ministry of Science, Technology and Innovation, Malaysia.

IOP Conf. Series: Materials Science and Engineering **716** (2020) 012015 doi:10.1088/1757-899X/716/1/012015

Carbonic anhydrase (CA) activity by *Chlorella* sp. in immobilised matrix under carbon dioxide rich cultivation condition.

M A Kassim[1]*, [1]M F I M Adnan, [1]K M Tan, M H A Bakar[1], J Lalung[1], M S Mohamed[2]

[1]Bioprocess Technology Division, School of Industrial Technology,
University Sains Malaysia (USM),11800 Penang, Malaysia.
[2]Faculty of Biotechnology and Biomolecular Sciences,
University Putra Malaysia (UPM)

*Email: asyrafkassim@usm.my

Abstract: The continuous release of global CO_2 and greenhouse gases into the atmophere is considered one of the major contributors for global warming. Currently, microalgal biosequestration using enzyme carbonic anhydrase (CA) has been reported to be one approach that could be applied to overcome the issue. Eventhough this enzyme has been proven to show its potential to convert atmospheric CO_2 to bicarbonates, there are remaining issues related to its stability and production parameters that need to be addressed. In this study, the activity of CA produced by immobilized microalgae *Chlorella* sp. cultivated in a laboratory environment was investigated. For this study, the influence of cultivation conditions such as pH value ranging from 4.00 to 12.00, light intensity ranging from 330 lux to 1000 lux and CO_2 concentration ranging from 0.04% to 25% on CA activity were investigated. This present study indicates that the highest CA activity of 1.908 U/min was observed for the cultivation was performed using 15% CO_2 with a pH of 8.00 and a light intensity of 550 lux. This suggested that the entrapment of microalgal using a suitable matrix carrier could produce higher CA activity which could be further utilised for extended biomimetic CO_2 capture systems.

1. Introduction

Carbon dioxide (CO_2) emissions are one of the major environmental problems generated from human activity. Over the years, a subtantial amount of CO_2 gas is emitted from various activity such as; transportation, animal, energy and the coal industry. Gas produced from these sources heat up the earth and enhance the global warming. Several carbon capture and storage (CCS) technologies that have captured CO_2 in the atmosphere and have been introduced to overcome this problem [1]. Technologies such as; physical processes (biochar burial, ocean storage, geological sequestration), chemical processes (mineral carbonization and chemical scrubber) and biological processes (reforestation, agriculture, photosynthetic microorganism) have been reported to have the potential to be applied in CCS [2].

Microbial CO_2 bio-sequestration is one of the most environmental friendly and cost effective methods to counteract CO_2 emissions [3]. Alternative approaches involve the utilization of CO_2 by photosynthesis microorganisms such as microalgae. Microalgae are unicellular groups that can be used to fix CO_2 emissions efficiently from different sources including the atmosphere, industrial exhaust gases, and soluble carbonate salts. These microorganisms are capable of utilizing CO_2 for its growth and at the same time producing a various valuable product [4, 5]. The benefits of using microalgae for environmental purposes are related to their growth rates and their tolerances regarding higher concentrations of CO_2.

The capability of microalgae to fix CO_2 is highly dependent on the special carbonic anhydrase (CA) enzyme which is key enzyme of the carbon concentrating mechanism (CCM) in algae [6, 7]. In general, the present of CO_2 concentration as an inorganic carbon in cultivation medium induces intra and extracellular carbonic anhydrase enzyme activity. The CO_2 essentials via the cell is transported into the cell by a hydration response depending upon the bicarbonate, CO_2 concentration inside and outside the cell. The transportation of CO_2 is through the transformation of carbon dioxide into bicarbonate and happens repeatedly in the cell. As a result, activation of CA activity under specific conditions increases the conversion rate of bicarbonate ions ($HCO3^-$) into CO_2 and subsequently promotes the uptake of CO_2 by the active transport into the microalgae cells [6]. According to Zeng et al., [8], microalgae have the ability to capture carbon dioxide gas, CO_2, in a single process and produce various valuable products at the same time.

The microbial carbon capture using CA from a free cell has been reported to show disadvantages such as difficult during its harvesting. Harvesting of the microalgae biomass, such as concentrating microscopic microalgae cells from dilute solutions of mass cultivation is an crucial step to secure high-quality effluents avoiding cell washout [9, 10]. On the other hand, in order to ensure the efficient CO_2 sequestration by microalgal CA, it is important to ensure the stability of enzymes and the microalgal growth. Therefore, immobilization of microalgae cells has been suggested to avoid harvesting issues. To date, there is little information on intracellular CAs activity produced by microalgae under immobilized conditions. Thus, this study is designed to investigate the CAs activity by *Chlorella* sp. in immobilized form cultivated under a CO_2 rich medium.

2. Material and Method

2.1 *Microalgae strain and cultivation*

Freshwater *Chlorella* sp. was used throughout the experiment. The culture was cultivated in an illuminator room. The mix was 0.3 L/min aired under light photons with an intensity of 1000 lux at 28-30C. The cultivation medium for the cultivation and seed preparation in this study consisted of 0.49 g/L magnesium sulfate ($MgSO_4.7H_2O$), 1.7 g/L sodium nitrate ($NaNO_3$), 0.14 g/L di-potassium phosphate (K_2HPO_4), and 0.03 g/L calcium chloride ($CaCl_2.2H_2O$) respectively.The growth of the culture was monitored daily by measuring its optical density and harvested during its late logarithmic growth phase prior to being subjected for the immobilization process.

2.2 *Immobilization*

The immobilization of *Chlorella* sp. was carried out using alginate as a immobilize matrix carrier. An active *Chlorella* sp. seed culture with an optical density of 680 nm (OD_{680}) of 1.0 was harvested by centrifugation at low speed (4500 rpm) for 10 minutes. Pellets rich with algal biomass was then washed and re-suspended in deionized water to form a concentrated algal suspension. The algal suspension was then mixed with a 4% sodium alginate solution in a 1:1 ratio in 4 mL of concentrated algal and a 4 mL sodium alginate solution. The mixture was gently stirred until the algae was evenly distributed. The mixture was dropped into 6% calcium chloride using a syringe to form uniform algal beads. The beads were left in a crosslink solution ($CaCl_2$) for approximately 12 h for hardening. After the hardening was complete, beads were separated from the solution using a sterile strainer/filter. The beads were rinsed several times with deionized water to remove any remaining calcium chloride. Blank alginate beads were prepared in the same manner as the algal beads with the exception of using deionized water rather than a cell suspension.

2.3 Preliminary cultivation of carbonic anhydrase analysis

A study on the effects of cultivation including pH, light intensity and CO_2 concentration on CA activity by *Chlorella* sp. in a immobilised matrix was initially investigated using a one-variable at a time design (OVAT) method. The effect of CA activity at pH range (4 to 12), light intensity range 300 to 1000 lux and CO_2 concentration range 0.04 to 25% (v/v) was assessed in this study. After a cultivation period, the CA activity in a immobilised matrix was analyzed as discussed in the next section.

2.4 Response surface methodology

A Box Behnken Design (BBD) was used in order to determine the interactive effects of cultivation parameters such as pH, light intensity and CO_2 concentration regarding CA activity. A multi-level three parameter matrix was employed to determine the synergistic effects between three parameters and the optimal point value. This was analyzed by fitting the second order polynomial model. The statistical significance of each parameter was determined through P values, where smaller p-values indicate effective parameters. The contour plots were used to show the individual and cumulative effects of variables of CA activity. These graphical representation models were plotted as a function of two variables while keeping other variables at the central level.

2.5 Determination of carbonic anhydrase activity

CA activity was determined using a method describe by Komala and Kiium [11]. Approximately 100 microalgal beads were crushed using a mortar then centrifuged at 200 xg for 10 minutes. The sample was suspended with a 0.1 phosphate buffer. The enzyme activity was determined using a spectrophotometric method using a 0.2 mL extracted sample mixed with 1.0 mL 3mM p-nitrophenyl acetate and a 1.8 and 0.1 M phosphate buffer. One sample of the solution was taken and used to measured the absorbance reading using a spectrophotometer at 348 nm. The changes of the absorbance at 348nm was recorded over the first 5 minutes in order to estimate the amount of the p-nitrophenyl (p-NP) released. The non-enzymatic hydrolysis was substracted by using a blank without enzymes. One unit of the enzyme activity represents the amount of enzyme catalyzing to produce 1 unit of p-nitrophenol per minute under the assay conditions. All the processes were done in duplicate.

3. Results and Discussion

Production of CA and its activity by microalgae could be influence by several factors. In this study, the effects of pH, light intensity and CO_2 concentration of CA production and its activity by microalgae in a immobilised matrix were investigated.

3.1Effect of pH

Figure 1 shows the effects of a pH value of cultivation medium on *Chlorella* sp.-CA activity in an immobilized matrix. Based on this study, the maximum CA activity of 1.02 U/min was obtained from the cultivation in a medium with an initial pH value of 7. It was found that a further increase of pH beyond pH 7 exhibited low CA activity.

It is known that pH value plays an important role in maintaining equilibrium of cultivation medium environments. Significant changes of pH values from 7 to 12 have reported to show a noteable affects on CA activity. Similar observations were reported on CA activity of *B. mucilaginosus* KO2 exhibiting maximum CA activity at pH 5.5 to 6.6 [12]. This finding also has been supported by a previous study reporting that typical maximum CA activity can be obtained at pH range 5 to 10 [13]. Previous studies reported that in some cases, CA activity by *Chlorella* sp. induces when the cultivation was perfomed at alkaline conditions [14]. A study by Vaquero et al. [15] reported that total CA activity by *Cocomyxa* sp. increased with at higher pH values with a maxiumym CA activity obtained when the cultivation was carried out using pH 6. Under alkaline conditions, the cultivation medium was predominantly carbon in the form of bicarbonate. This required CA transport for this compund into a membrane cell [16]. This present study also showed that the microalgal could assist CO_2 sequestration via CA at this pH range which proves very usefull application which can be applied for CO_2 capture and alkaline wastewater treatment.

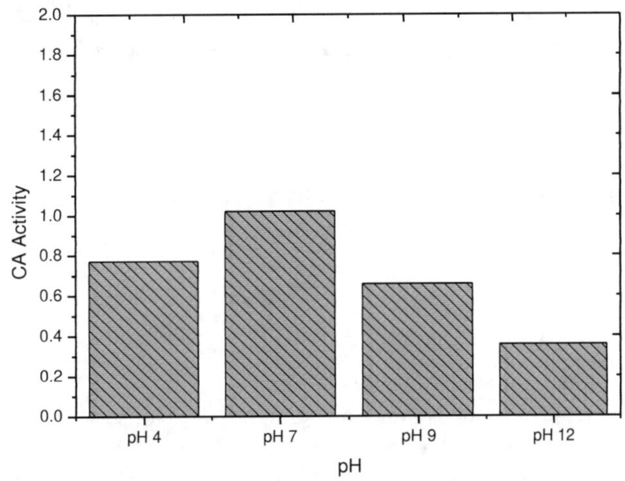

Figure 1. Effect of initial pH value on the CA activity in immobilised *Chlorella* sp.

3.2 *Effect of light intensity*

The influence of light intensity on the CA activity of *Chlorella* sp. in alginate polymer was also studied. Results are shown in Figure 2. The present study showed that exposure of immobilised *Chlorella* sp. could significantly influence the the CA activity. The maximum CA activity of 1.021 U/min was found when the cultivation was carried out using 550 lux. Further increase of light intensity and cultivation using low light could reduce CA activity. Results obtained from this study is in agreement with previous reports that suggest exposure of microalgae to higher light intensities could induce CA activity [17, 18]. Similar observations has also been reported on CA activity of *Tetraselmis gracilis*, in which 50% of CA activity loss was detected when cultivation was performed under low light conditions [19].

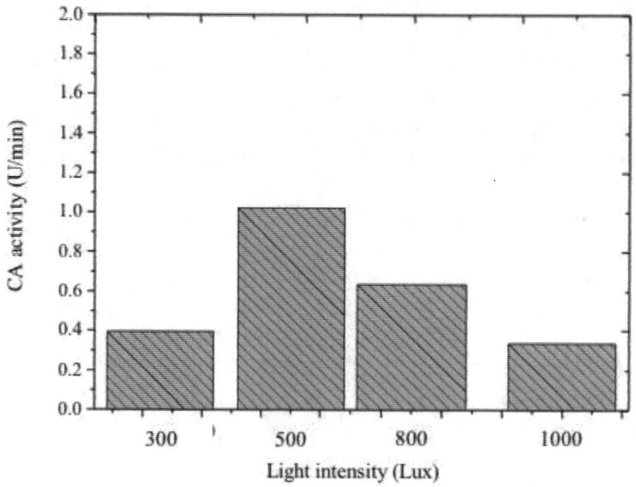

Figure 2. Effect of light intensity on the CA activity in immobilised *Chlorella* sp.

3.3 Effect of CO_2

CO_2 concentration plays an important role in microalgae growth and its carbon capture activity [20, 21]. In this study, the effects of CO_2 concentration on CA activity in an immobilised matrix was evaluated using CO_2 concentration ranging from 0.04 to 25% (v/v). Results are presented in Figure 3. Based on this study, it was clearly indicated that higher CA activity was observed for the cultivation using a higher CO_2 concentration compared to normal air. It was found that maximum CA activity in an immobilised matrix of 1.721 U/min can be obtained when the cultivation was performed using 15% CO_2. An increase of CO_2 concentration from normal air to 15% CO_2 could enhance the CA activity up to 41%. However, cultivation of microalgae using CO_2 concentrations up to 25% CO_2 could reduce CA activity.

Varied observations have been reported on the effects of CA activity by different microalgae species [19, 22]. Studies indicate that CA activity by *Tetraselmis gracilis* increased when the CO_2 level decreased suggesting that CO_2 was used by the microalgal via assitant of the CA enzyme. According to Xia et al. [23], reports that higher CA activity by *Chlorella pyrenoidosa* and *Chlamydomonas reihardtii* were observed when the cultivation was perfomed at low CO_2 concentrations. In contrast, CA activity by *Scenedesmus obliqus* cultivated using different CO_2 concentrations showed the highest CA activity when this strain was supplied with 35% CO_2 after 15 days of cultivation [24]. High CA activity reported in this study represents internal and external CA activity. These results therefore imply that exposure to suitable CO_2 concentrations could favour CA activity by *Chlorella* sp. in a immobilised matrix. Differences in CA activity reported in this study also showed that CA activity by the microalgae is species dependent.

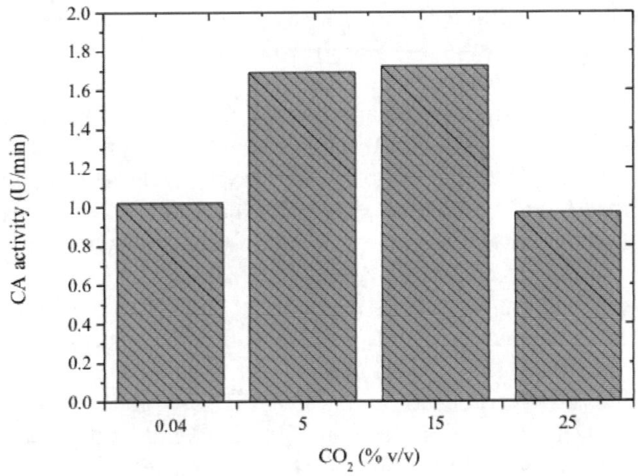

Figure 3. Effect of CO_2 concentration on the CA activity in immobilised *Chlorella* sp.

3.4 Effect of cultivation on enzyme activity

Further studies on the effects of cultivation conditions such as pH, light intensity, and CO_2 concentration for CA activity were carried out using the Box Behnken response surface design (BBD). Based on the present study, the quadratic model was found to be the most suitable explanation to for the interaction of cultivation parameters on CA activity in alginate polymer with P > F value of < 0.05 (Table 1). The coefficient of determination (R^2) of the regression equation for CA activity was 0.97 (Adj = 0.92). The R^2 value was close to 1 indicating that the predictability of the value is at 95% confidence level and the response function for the predicted value agreed well with the experimental data.

Table 1: Analysis of variance (ANOVA) for CA activity by immobilized *Chlorella* sp.

Source	**SS**	**Df**	**MS**	**F value**	**Prob > F**
Model	5.08	9	0.56	6.60	0.0105
CO_2	0.25	1	0.25	2.95	0.1296
pH	0.12	1	0.12	1.42	0.2718
Light intensity (LI)	0.38	1	0.38	4.44	0.0730
CO_2 x pH	0.28	1	0.28	3.31	0.1118
CO_2 x LI	0.22	1	0.22	2.53	0.1556
pH x LI	0.029	1	0.029	0.34	0.5761
CO_2x CO_2	2.52	1	2.52	20.42	0.0010
pH x pH	0.27	1	0.27	3.14	0.1195
LI x LI	0.70	1	0.70	8.21	0.0241
Residual	0.60	7	0.086		
Lack of fit	0.44	3	0.15	3.79	0.1152
Pure error	0.16	4	0.039		
R2	0.9654				
Adj-R2	0.9210				

DF: degree of freedom of variance source, SS: sum of squares, MS: mean of squares.

A three dimensional (3D) response surface plot analysis was performed to determine the interaction and the most suitable condition for total CA activity in the immobilised matrix with the response of chosen parameters. Figure 4 shows the 3D contour plots for the effects of CO_2 concentration (%) and initial pH on carbonic anhydrase (CA) activity (U/min). From this figure, it shows that the increase of CO_2 concentration (%) and pH will increase CA activity. Further cultivation beyond CO_2 concentration and pH will reduce CA activity of *Chlorella* sp. in an immobilized matrix. The decreasing of CA activity was due to the use of a higher CO_2 level resulting the low in pH. This is because the microalgae *Chlorella* sp. showed maximum biomass productivity at pH 6.5. It has been reported that the decreasing of pH will reduce the activity of CA and inhibit cell growth. Therefore, both interactions play important roles to the production of CA activity.

Figure 4B reveals the 3D contour plots and the interaction effects of CO_2 concentration (%) and light intensity (lux) on carbonic anhydrase (CA) activity (U/min). The increase of CO_2 concentration (%) from 0.04% to 15% significantly increased CA activity. However, further cultivation beyond this value slowly reduced CA activity. For various light intensities (lux), the cultivation of high light intensity from 330 lux to 550 lux significantly increased CA activity. However, further cultivation beyond this value slightly decreased CA activity. According to Solovchenko et al., [25], the higher the light intensity the more reduction in growth rate resulting damage to the cells. Previous studies found that microalgae showed higher activity and growth at a light intensity of 3000 lux but the evaporation rate of the medium increased and most of the cells dried in the flask when illuminated to direct sunlight. Therefore, selecting a suitable CO_2 concentration and light intensity is important in order to produce high CA activity.

Figure 4C shows the interaction of an initial pH and light intensity (lux) on carbonic anhydrase (CA) activity (U/min) by microalgae in the alginate matrix. The present study indicated that the increase in pH value from 4.00 to 8.00 increases CA activity. Further cultivation using a high alkaline medium with a pH value of 12.00 slightly reduced CA activity. Generally, the maximum CA activity from microalgae is induced at alkaline condition. Previous study has reported that the maximum extracellular CA activity of *Chlorella* sp. is at pH 7.5 and 8.5 [26]. For the light intensity (lux), an increase of light intensity from 330 to 550 lux increasec CA activity. However, for the cultivation of a high light intensity slowly decreased CA activity. Therefore, the interaction between pH and light intensity did not really effect the production of CA activity.

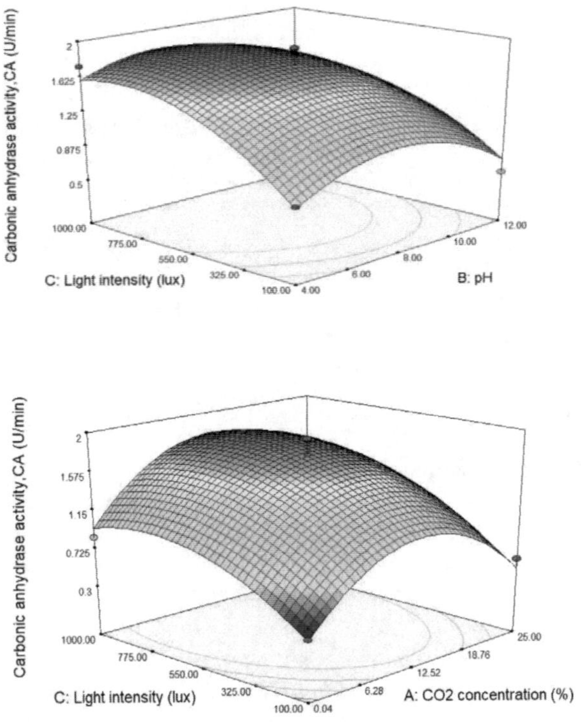

Figure 4. 3D plot on the effect of cultivation condition on the CA activity (a) effect of pH and light intensity (b) effect of pH and CO_2 concentration and (c) effect of light intensity and CO_2 concentration

4. Conclusion

In conclusion, this work showed that the cultivation condition parameters including the initial cultivation pH, light intensity and CO_2 are important variables that may be required to produce maximum CA activity by *Chlorella* sp. in an alginate polymer as a immobilised matrix carrier. The maximum CA activity of 1.908 U/min was observed when the cultivation was conducted using 15% of CO2 , with a pH of 8.00 and a light intensity of 550 lux. This present study also demonstrated that a microalgae biomass has great potential as a source of a carbonic anhydrase enzyme. The information generated from this study is crucial to determine the most suitable strategy for to be applied for CO_2 sequestration using CA enzyme.Adapting the most stable condition could play an important role to ensure the reduction of CO_2 and keeping environmental sustainability. On the other hand, results generated could assist in understanding the biomineralization by microalgae under an intergrated cultivation system, especially involves indutsrial processing plant.

References

[1] Araújo OdQF, and de Medeiros JL 2017 Carbon capture and storage technologies: present scenario and drivers of innovation *Current Opinion in Chemical Engineering*, **17** 22- 34.

[2] Grover M 2015 Elevated CO2: Plant associated microorganisms and carbon sequestration *Applied Soil Ecology* **95** 73-85.

[3] Prabhu C, 2009 Immobilization of carbonic anhydrase enriched microorganism on biopolymer based materials *Journal of Molecular Catalysis B: Enzymatic* **60** 13-21.

[4] Kassim MA and Meng TK 2017 Carbon dioxide (CO_2) biofixation by microalgae and its potential for biorefinery and biofuel production *Science of The Total Environment* **584-585** 1121-1129.

[5] Chen CY, Kao PC, Tsai CJ, Lee DJ, Chang JS 2013 Engineering strategies for simultaneous enhancement of C-phycocyanin production and CO_2 fixation with *Spirulina platensis Bioresource Technology* **145** 307-312.

[6] Tang D, Han W, Li P, Miao X, Zhong J 2011 CO_2 biofixation and fatty acid composition of *Scenedesmus obliquus* and *Chlorella pyrenoidosa* in response to different CO_2 levels. *Bioresource Technology* **102** 3071-3076.

[7]Boone CD, Gill S, Habibzadegan A, McKenna R 2013 Carbonic anhydrase: An efficient enzyme with possible global implications *International Journal of Chemical Engineering* **2013**: p. 6.

[8] Zeng, X., Danquah MK, Zhang S, Zhang X, Wu M, Chen XD, Ng IS, Jing K, Lu Y 2012 Autotrophic cultivation of *Spirulina platensis* for CO_2 fixation and phycocyanin production *Chemical Engineering Journal* **183** 192-197.

[9] Richmond A, Cheng-Wu Z, Zarmi Y 2003 Efficient use of strong light for high photosynthetic productivity: interrelationships between the optical path, the optimal population density and cell-growth inhibition *Biomolecular Engineering* **20**(4) 229-236.

[10] Muñoz R, and Guieysse, B 2006 Algal–bacterial processes for the treatment of hazardous contaminants: A review *Water Research* **40**(15) 2799-2815.

[11] Komala T, and Khun TC 2014 Biological carbon dioxide sequestration potential of *Bacillus pumilus.* Sains Malaysiana **43** 1149-1156.

[12] Zhang Z, Lian B, Hou W, Chen M, Li X, Shen W and Li Y 2011 Optimization of nutritional constituents for carbonic anhydrase production by *Bacillus mucilaginosus* K02 *African Journal of Biotechnology* **10**(42) 8403-8413.

[13] Beuf L, Kurano N, and Miyachi S 2000 Effect of external pH on inorganic carbon assimilation in unicellular marine green algae *Phycological Research* **48**(1) 47-54.

[14] Vaquero I, Vazquez M, Ruiz-Dominguez MC, Vilchez C 2014 Enhanced production of a lutein-rich acidic environment microalga *Journal of Applied Microbiology* **116**(4) 839-850.

[15] Shiraiwa Y, and Miyachi S, 1985 Effects of temperature and CO_2 concentration on induction of carbonic anhydrase and changes in efficiency of photosynthesis in *Chlorella vulgaris* 11h *Plant and Cell Physiology* **26**(3) 543-549.

[16] Ores, JC, De Amarante MCA, Fernandes SS, Kalil SJ 2016 Production of carbonic anhydrase by marine and freshwater microalgae *Biocatalysis and Biotransformation* **34**(2) 57-65.

[17] Dionisio ML, Tsuzuki, M, and Miyachi S 1989 Light requirement for carbonic anhydrase induction in *Chlamydomonas reinhardtii Plant and Cell Physiology* **30**(2) 207-213.

[18] Borodin VB, 2008 Effect of red and blue light on acclimation of *Chlamydomonas reinhardtii* to CO_2-limiting conditions *Russian Journal of Plant Physiology* **55**(4) 441.

[19] Rigobello-Masini M, Aidar E, and Masini JC 2003 Extra and intracelular activities of carbonic anhydrase of the marine microalga *Tetraselmis gracilis* (Chlorophyta) *Brazilian Journal of Microbiology* **34** 267-272.
[20] Almomani F, Ketife AA, Judd S, Shurair M, Bhosale RR, Znad H, Tawalbeh 2019 Impact of CO_2 concentration and ambient conditions on microalgal growth and nutrient removal from wastewater by a photobioreactor. *Science of The Total Environment* 662 662-671.
[21] Neves FdF, Hoinaski L, Rorig LR, Derner RB, Lisboa HdM 2018 Carbon biofixation and lipid composition of an acidophilic microalga cultivated on treated wastewater supplied with different CO_2 levels. *Environmental Technology* **40**(25) 3308-3317.
[22] Nimer NA, Brownlee C, and Merrett MJ 1999 Extracellular carbonic anhydrase facilitates carbon dioxide availability for photosynthesis in the marine dinoflagellate *Prorocentrum micans Plant physiology* **120**(1) 105-112.
[23] Xia JR, and Gao KS 2005 Impacts of elevated CO_2 concentration on biochemical composition, carbonic anhydrase, and nitrate reductase activity of freshwater green algae *Journal of Integrative Plant Biology* **47**(6) 668-675.
[24] Basu S, Roy AS, Mohanty K, Ghoshal AK 2014 CO_2 biofixation and carbonic anhydrase activity in *Scenedesmus obliquus* SA1 cultivated in large scale open system *Bioresource Technology* **164** 323-330.
[25] Solovchenko AE, Khozin-Golderg I, Didi-Cohen S, Cohen Z, Merzlyak MN 2008 Effects of light intensity and nitrogen starvation on growth, total fatty acids and arachidonic acid in the green microalga *Parietochloris incisa Journal of Applied Phycology* **20**(3) 245-251.
[26] Li L, Fu ML, Zhao YH, Zhu YT 2012 Characterization of carbonic anhydrase II from *Chlorella vulgaris* in bio-CO_2 capture *Environmental Science and Pollution Research* **19**(9) 4227-4232.

Acknowledgment
This work was supported by School of Industrial Technology, Universiti Sains Malaysia (USM) and USM Fellowship. The author also would like to thank Research University Insentive (RUI) Grant 1011/PTEKIND/8011043) for financial support.

Potential use of eco-enzyme for the treatment of metal based effluent

M. Hemalatha and P.Visantini

School of Biological sciences, Quest International University Perak, 30250, Ipoh, Perak, Malaysia

Email: hemalatha.murugaiah@gmail.com

Abstract. Conversion of wastes into value added products is potentially profitable besides creating a clean environment. Eco-enzyme is one such product produced from fermentation of citrus fruit wastes and the effectiveness of the enzyme in treating metal based effluent was investigated. The characteristics of effluent before and after treatment with eco-enzyme were analysed. There was a great reduction in the level of BOD from 80.0 mg/L to 22.3 mg/L. Eco-enzyme also inhibits the growth of microorganisms in the wastewater. These results confirmed that eco enzyme is capable to treat metal based effluent. Besides, sludge treated with eco-enzyme was studied for its potential use in enhancing the growth of chili and aloe vera. Soil containing eco-enzyme treated sludge promotes better growth in chili. Meanwhile, aloe vera grown on soil containing eco-enzyme treated sludge grown almost equally and healthy in comparison with control. This study provides a solution to prevent sludge dumping to ensure clean environment.

1. Introduction

In the era of global industrialization, environmental pollution has become critical due to the increase in the worldwide population. This eventually generates more food materials especially fruits, vegetables and cereals to meet the demand. The total production of fruits and vegetables in the world is in increasing trend and at the same time, around 30-40% of the total production is being eliminated as waste due to various reasons [1]. Attempts are being made to protect the environment by recycling the wastes and make use of them for value added applications. Residues of citrus peels can be used for production of various enzymes using fermentation process, where the waste is converted into value-added products [2]. These citrus wastes can be used to produce eco-enzymes by undergoing fermentation process for 3 months after being mixed with brown sugar and water [2]. Eco-enzyme also refers as garbage enzyme is a multi-enzyme solution comprising protease, lipase, and amylase [3]. This enzyme was invented by a researcher in Thailand in the year 2006 by using organic solid waste. Eco-enzyme which is a vinegar-like liquid will be formed after three months of fermentation process using organic solid waste and can serve as a potential solution to treat waste water [4].

Industrialization and innovation have put load on the nature by releasing massive amount of toxic waste such as heavy metals, metalloids and organic contaminants. This caused severe impact on the nature. Myriad chemical and biological ways are available to treat the industrial waste, yet the problem is not solved due to high cost and longer duration taken to effectively treat the waste. Demand for wastewater treatment plants in industrial and municipal wastewater are increasing rapidly to comply with the allowed limit of wastewater to be discharged in environment [4]. Treating wastewater become a challenging task due to the over production of sludge in industries. Incineration and landfilling are the most common methods which are used to treat sludge [5]. However, the developing nations came up with new laws which force the industries to minimize the amount of sludge to be released to the environment and recommended to contemplate alternative method to reuse the sludge. This study was conducted to treat the industrial effluent rich in metals from electronic industry in Ipoh, Perak by using eco-enzyme which have been produced using citrus fruit peels. Subsequently, potential use of eco-enzyme treated sludge for plant growth was also investigated.

2. Materials and Methods

2.1. Production of eco-enzyme
Eco - enzyme was produced by adding sugar, citrus waste and water in the ratio of 1:3:10 left to undergo fermentation process for three months [3]. The enzyme produced was filtered and stored at room temperature for further analysis.

2.2 Sampling and characterization of industrial effluent
Effluent was collected from electronic based industry (Latitude; 4.6499° N, Longitude; 101.0730° E) in Ipoh and characterized based on these parameters; Biological oxygen demand (BOD), Total solid (TS), Total dissolved solid (TDS), Total suspended solid (TSS), pH, and colony forming unit (C.F.U). BOD of samples was measured using a probe (Fisher Scientific™ accumet™ waterproof AP84 Portable Dissolved Oxygen Meter, Singapore). Evaluation of TS, TDS and TSS of samples were conducted according to Standard Methods published by Public health association, American water works association, water environment federation.

2.3 Treatment of effluent using eco-enzyme
Eco-enzyme was used to treat the wastewater and sludge. Treatment was conducted for 5 days. Characteristics of treated wastewater and sludge was analysed based on pH, BOD, TS, TSS, TDS, and C.F.U. to achieve the Standard A water quality set by the United States Environmental Protection Agency (EPA).

2.4 Cultivation of plants using eco-enzyme treated sludge
Chili (*Capsicum annuum*) and aloe vera (*Aloe vera*) were used to study the application of eco-enzyme treated sludge for plant growth. Growth medium containing soil treated with eco-enzyme was prepared. Another medium containing soil and eco-enzyme treated sludge in the ratio of 1:1 was also prepared. The physical changes of the plants under treatments were observed and recorded for the period of 10 weeks. Plants were analysed based on height and the condition of the plants. The data was subjected to analysis of variance (ANOVA) and samples means tested for significant differences using statistical analysis (IBM SPSS Statistics 20).

3. Results

3.1 Characterization of eco-enzyme treated industrial effluent
Eco-enzyme produced after 3 months of fermentation was filtered and used to treat the wastewater collected from metal based industry. Characteristics of effluent before and after treatment with eco-enzyme were analysed and indicated in Table 1. The wastewater was treated with eco-enzyme for the duration of 5 days. It was proven that, eco-enzyme has a promising effect in treating the wastewater as

the reduction in TS, TDS and TSS was observed. Concurrently, the level of BOD significantly ($P<0.05$) reduced from 80 mg/L to 22.3 mg/L in effluent. Besides, when the wastewater treated with eco-enzyme, there was no bacterial growth observed. Various hydrolytic enzymes present in eco-enzyme increase the accessibility of the enzyme to penetrate the bacterial cell. This will eventually lead to the inhibitory effect against bacterial growth. Besides, eco-enzyme is acidic in nature and lead to the lytic effect of the microorganisms tested [4][6].Therefore this findings supports the fact that eco-enzyme has antimicrobial property. Overall, eco-enzyme offers an alternative to chemical treatment since EPA standard was met after the treatment using eco-enzyme. Wastewater treatment of eco-enzyme takes longer period of time, but it is safe to be used instead of carcinogenic chemical such as Ethylenediaminetetraacetic acid (EDTA) [7] which in turn could be dangerous to workers who handling it despite affecting the environment.

Table1 Characteristics of eco-enzyme treated wastewater

Parameters	**Effluent**	
	Control	**After treatment**
pH	7.4	7
TS (mg/ml)	464.2	313.5
TDS (mg/ml)	365	221.3
TSS (mg/ml)	97.0	65.0
BOD (mg/L)	80	22.3
c.f.u ml^{-1}	1.5×10^{10}	0

3.2 Study on the potential application of eco-enzyme for plant growth

Effect of eco-enzyme in stimulating plant growth using sludge was investigated. Sludge which was treated with 25% of eco-enzyme used to grow chili and aloe vera for the duration of 10 weeks. Eco-enzyme was diluted prior to use as concentrated eco-enzyme can cause the soil to become acidic which eventually affects the growth of plants [8]. It was observed that the soil containing eco-enzyme treated sludge significantly ($P<0.05$) promote growth of chili plant in comparison with the medium containing only soil and soil with eco-enzyme (Table 2). The height of chili plant on week 10 was 12.8±1.1 cm. Meanwhile, the height of control was only 2.9±0.19 cm. The conditions of the chili plants after 10 weeks of growth are indicated in Figure 1. Sludge is rich with organic matter and other nutrients [9]. This is a possible reason, when soil with eco-enzyme treated sludge was used; significantly higher growth was recorded in chili plant.

Table 3 shows the length of aloe vera under various growth medium. There were no significant changes in the length of leave in aloe vera grown for the duration of 10 weeks. Interestingly, growth medium containing sludge did not supress the growth of aloe vera. The plant grown on soil treated with eco-enzyme and soil treated with eco-enzyme treated sludge comparatively show healthier growth than control on week 10 (Figure 2). Eco-enzyme containing multiple hydrolytic enzymes may have significant effect in maintaining soil health [10] which eventually promotes the growth of chili and alove vera. These promising results offer an alternative to utilize the sludge for agriculture and eventually offer a solution to reuse the sludge in waste water treatment. These findings also propose a solution to the issue of sludge dumping for sustainable environment.

Table 2 Effect of eco-enzyme on the growth of chili under various medium

Duration (weeks)	Height of plant (cm)		
	Soil (Control)	Soil treated with eco-enzyme	Soil with eco-enzyme treated sludge
2	2.30±0.1[a]	3.60±0.1[b]	4.20±0.2[c]
4	2.50±0.1[a]	4.20±0.2[b]	5.50±0.2[c]
6	2.50±0.2[a]	4.50±0.2[b]	7.40±0.4[c]
8	2.80±0.2[a]	4.70±0.2[b]	9.50±0.3[c]
10	2.90±0.2[a]	5.00±0.2[b]	12.8±1.1[c]

*Values are mean±standard deviation triplicates. Mean with different supergenote ([a.b.c]) within the same column are significantly different at $p<0.05$

Figure 1 Condition of chili plant on week 10 in growth medium containing (a) Soil (Control), (b) Soil treated with eco-enzyme, (c) Soil with eco-enzyme treated sludge

Table 3 Effect of eco-enzyme on the growth of aloe vera under various medium

Duration (weeks)	Height of plant (cm)		
	Soil (Control)	Soil treated with eco-enzyme	Soil with eco-enzyme treated sludge
2	15.1±0.2[a]	15.2±0.2[a]	15.3±0.3[a]
4	15.5±0.3[a]	16.0±0.2[a]	16.3±0.4[a]
6	16.0±0.5[a]	17.0±0.5[a]	17.2±0.4[a]
8	16.5±0.4[a]	17.8±0.3[a]	18.0±0.5[a]
10	17.2±0.7[a]	18.6±0.5[a]	19.0±0.6[a]

*Values are mean±standard deviation triplicates. Mean with different supergenote ([a.b.c]) within the same column are significantly different at $p<0.05$

Figure 2 Condition of aloe vera on week 10 in growth medium containing (a) Soil (Control), (b) Soil treated with eco-enzyme, (c) Soil with eco-enzyme treated sludge

Conclusion

In the present study eco-enzyme produced using citrus fruit waste offer potential solution to treat waste water in eco-friendly way. About 70% reduction in the level of BOD from metal based effluent was observed. Effectiveness of eco-enzyme in treating waste water was proven based on the reduction obtained for TS (32.5%), TDS (39.5%) and TSS (33.0%). It is suggested that eco-enzyme can also inhibit the growth of microbes which was proven based on the C.F.U value obtained. Study on the applications of eco-enzyme to improve soil fertility reveals that chili and alove vera showed improved growth in the duration of observation (10 weeks). Chili showed significant growth when the growth medium containing soil and sludge treated with eco-enzyme was used. Aloe vera showed a promising growth in the presence of eco-enzyme compared to control which contain only soil. This results prove that eco-enzyme improve the fertility of the soil. Besides, when the sludge treated with eco-enzyme, enhanced growth of chili and aloe vera were also recorded. Although it need further study on the mechanism of the eco-enzyme in treating waste water, this research offer a potential solution to treat metal based effluent and suggest a prospective use of sludge in agriculture for sustainable environment.

References

[1] Pleissner D and Lin C S K 2013 Valorisation of food waste in biotechnological processes *Sustainable Chem Processes* **1(1)** 21-6.

[2] Arun C and Sivasanmugam P 2017 Study on optimization of process parameters for enhancing the multi-hydrolytic enzyme activity in garbage enzyme produced from preconsumer organic waste *Biores Tech* **226** 200-10

[3] Tang FE and Tong CW 2011 A study of the garbage enzyme's effects in domestic waste water *World Academy of Eng and Tech* **60** 1143-48

[4] Arun C and Sivasanmugam P 2015 Investigation of biocatalytic potential of garbage enzyme and its influence on stabilization of industrial waste activated sludge *Process Safety and Environ Protec* **9** 471-8

[5] Gaur N Flora G Yadav M and Tiwari A 2014 A review with recent advancements on bioremediation-based abolition of heavy metals *Environ Sci: Processes & Impacts* **16(2)** 180-93

[6] Straub TM, Pepper IL and Gerba GP 1993 Hazards from pathogenic micro-organisms in land-disposed sewage sludge *Rev of Environ Contam and Toxic* **132** 55-91

[7] Awaleh MO and Soubaneh YD 2014 Waste water treatment in chemical industries: The concept and current technologies *Hydrology Curr Res* **5(1)**

[8] Marschner B and Noble A D 2000 Chemical and biological process leading to the neutralisation of

acidity in soil incubated with litter materials *Soil Bio and Biochem* **32** 805-13
[9] Tao J, Wu S, Sun L, Tan X, Yu S, Zhang Z 2012 Composition of waste sludge from municipal wastewater treatment plant *Procedia Environ Sci* **12** 964-971
[10] Das SK and Varma A 2011 *Soil Enzymology* Springer-Verlag Berlin Heidelberg

Fructose stearate esterify in packed bed reactor using immobilized lipase.

N A Shahrin, G G Yong and N A Serri

Bioprocess Technology, School of Industrial Technology, Universiti Sains Malaysia, 11800 Penang, Malaysia

Email: aziah_serri@usm.my

Abstract . The enzymatic esterification of sugar-fatty acid ester to produce bio-based surfactants or emulsifiers has been recognized as an alternative way to the chemical synthesis due to its environmentally friendly reaction. Therefore, present study aimed to employ an optimal procedure for the continuous synthesis of fructose stearate by using immobilised *Rhizomucor miehei* lipase (RML) in a packed-bed reactor. Briefly, lipase immobilization on chicken eggshells was conducted and characterized using Transmission Electron Microscopy (TEM) and Brunauer-Emmett-Teller (BET) analysis. Subsequently, the screening of enzyme loading was performed. Response Surface Methodology (RSM) based on central composite design (CCD) was applied to optimize the temperature, flow rate and substrate molar ratio. The immobilisation efficiency on eggshells was 63.64%. After immobilization, the BET surface area, total pore volume and pore diameter of the eggshells were reduced to 1.0714 m^2/g, 1.003×10^{-3} cm^3/g and 3.7449 nm, respectively. Furthermore, both BET analysis before and after immobilization revealed that the pore structures of eggshells were classified as Type II isotherm. From preliminary study, enzyme loading of 1.5 g immobilized lipase was selected as the optimum enzyme loading. The quadratic model in RSM analysis was validated to predict the optimum conditions. A maximum of fructose stearate concentration as high as 7.252×10^{-1} mol/L obtained at a better condition of 37.47^{0}C under a flow rate of 0.074 ml/min and 2.82:1 substrate molar ratio of fructose to stearic acid. This work has pronounced the eggshell is as a potential carrier for RML immobilization with ability to be used in packed bed reactor to synthesis fructose stearate.

1. Introduction

Sugar esters are attractive and biocompatible surfactants which possess excellent stability, detergency, foaming, emulsifying and dispersing effects, turning it into one of the most versatile process chemicals [1]. In synthesis of sugar ester, the lipase-catalyzed reaction theoretically inserts a fatty acid into a specific position of the sugar. Lipases have substrate specificity towards the acyl donors that are varying in carbon chain length. *Rhizomucor miehei* lipase (RML) is sn-1,3 specific lipase in which its sn-1,3 position is usually held by unsaturated fatty acids, such as oleic acid, palmitic acid and stearic acid [2]. Adsorption of lipase on hydrophobic eggshells support occurs surrounding the active center at lower ionic strength if compared to other enzymes. Lipase will recognize the supports as a lipid/water interface [3]. During hydrophobic adsorption, lipases are strongly absorbed on the interface of hydrophobic supports and switch the conformational equilibrium of lipase towards the stabilized open form [4]. Enzyme in its immobilized form has many favorable qualities, including enhanced stability towards extreme environment, convenient handling, ease separation of products, and efficient recovery and reuse [5-6].

In the present study, enzymatic esterification of fructose and stearic acid by immobilized RML was conducted in packed bed reactor (PBR). PBR enables the reaction to run at higher enzyme loading and shorter diffusion pathway than that of the batch conditions, leading to higher conversion yield and higher rate of reaction per unit of enzyme [7]. Furthermore, reaction under continuous mode via PBR can increase the automated operation and surmount a problem of the long-time reaction in batch approach due to large amount of downtime to be scheduled for emptying, cleaning, and filling [8].

Hence, this study was carried out to study the efficiency of immobilization approach using eggshells by means of surface and porosity values from Brunauer-Emmett-Teller (BET) method. The aim was also to obtain optimum conditions of temperature, flow rate and substrate molar ratio of fructose to stearic acid for enzymatic synthesis of fructose stearate using immobilized enzyme in packed bed reactor via response surface methodology (RSM).

2. Experimental

Chicken eggshells were collected from the hawker stall around Universiti Sains Malaysia (Penang, Malaysia). Commercial *Rhizomucor miehei* lipase (RML) was purchased from Sigma-Aldrich and other chemicals used throughout the study were purchased from Merck, QRec, R&M Chemical and Sigma-Aldrich.

2.1. Pre-treatment of chicken eggshells

Prior to lipase immobilization, the chicken eggshells were pre-treated by using the optimized conditions obtained from preliminary studies done by Salleh et al. (2016). All dried eggshell was sieved with 63 µm pore size.

2.2. Lipase immobilization into chicken eggshells

Using adsorption technique, RML was immobilized on pre-treated eggshell. About 67 % (v/v) of lipase in phosphate buffer (50 mM, pH 7) was loaded into the flask along with 10 g of eggshell. The immobilization was carried out at 37 °C and 200 rpm for 1 hour and later being stored at 4°C for 16 hours after incubation. Unbound lipase was removed by washing with phosphate buffer [9]. Immobilized lipase was checked for its enzyme activity using tributyrin method [3]. The size and morphology of eggshell before immobilization was determined by transmission electron microscopy (TEM). Surface characterization for both eggshell powder before and after immobilisation was carried out at liquid N2 temperature of 77K using Micromeritics ASAP 2020. The specific surface area of samples *SBET*, was determined using multipoint Brunauer-Emmett-Teller (BET) method. The Barrett- Joyner-Halenda (BJH) method was used to measure pores' volume, *Vpore* and the pore diameter distribution [10-12].

2.3. Continuous synthesis of fructose stearate using packed bed reactor

The PBR system which comprised of a cylindrical glass column (10 mm i.d.× 90 mm length) was filled with starting mixture of fructose and stearic acid well dissolved in ethanol (0.5 M to 1.0 M) in a feeding tank incorporated with immobilized RML. A re-circulating water bath was employed to
maintain the desired temperature and by-pass valve was required to lower the pressure of the reactor system, thus allow the substrate to flow back to the feed tank. Range of conditions (temperature, flow rate and substrates molar ratio of fructose to stearic acid) obtained from preliminary was further optimized using response surface methodology. Design Expert 7.0.0 software was used to design and analyze the experimental data. After factors screening in preliminary study, significant parameters were selected to undergo face centered central composite design (CCD) in response surface methodology to determine the optimum condition for producing the highest concentration of fructose stearate. The experimental sequence for a total of 15 experiment runs suggested by software was randomized to reduce the effects of uncontrolled factors.

IOP Conf. Series: Materials Science and Engineering **716** (2020) 012017 doi:10.1088/1757-899X/716/1/012017

2.4. High-performance liquid chromatography

The quantitative analysis of residual fructose concentration was performed by Shimadzu high performance liquid chromatography (HPLC) equipped with Thermo Scientific Hypersil™ APS-2 HPLC Columns with mobile phase mixture of acetonitrile and deionized water, 85:15 by volume at a flow rate of 1.0 ml/min. The sample volume of 20 µL was injected into the HPLC system and eluted at 35 °C. The concentration of fructose stearate was calculated as:

$$\text{Concentration of fructose stearate} = (X_0 - X_1)/X_0 \times 100 \qquad \text{Eq. 1}$$

where X_0 is the fructose concentration at the beginning of reaction; X_1 is the residual fructose concentration in the mixture at the end of the reaction.

One-way analysis of variance (ANOVA) was employed to evaluate the reliability and validity of the results obtained. The results are considered as statistically significant if the *p-value* < 0.05.

3. Results and discussion

The TEM images of eggshell powder before immobilization shown in Figure 1 illustrated that eggshell powder was made up of mixture of irregular rhombohedral and quasi-spherical shapes. Similar observation had been reported by Rahman *et al*., (2014) for eggshell nanopowder with the particle size in the range of 400 nm to 900 nm. The eggshell powder also exhibited its highly agglomerated nature that revealed many bright stripes, known as the ordered crystal lattice fringes which can be observed in various directions. In general, TEM analyses reflected high degree of crystallinity of eggshell powder. The results further support that the eggshell has high mechanical strength that makes it a suitable carrier in enzyme immobilization [12-14].

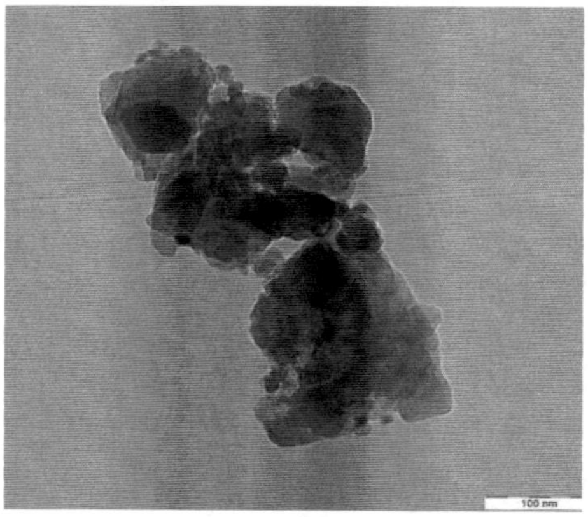

Figure 1 . TEM image of eggshell powder before immobilisation at 50 K magnification

From Table 1, BET analysis showed that surface area decreases from 2.1364 m^2/g before lipase immobilization to 1.0714 m2/g after immobilization. There was also a significant decrease in the average pore diameter of the eggshell after immobilization. Reduction of eggshell powder pore volume of eggshell also observed after immobilization. The huge reduction found in surface area, pore diameter and pore volume indicated that the enzyme was successfully adsorbed on eggshell in which was similar as the finding by Rodrigues *et al*., (2008) in the immobilization of *Candida antarctica* lipase on activated carbon [15]. From tributyrin assay conducted, the enzyme activity before and after immobilization were also determined and a high enzyme immobilization efficiency of 63.64% was obtained (result not shown). Upon immobilization, RML filled up some spaces of the pores and destroyed other pores simultaneously, constructing new pores with various sizes and irregular shapes, resulting in the decrease in surface area, pore diameter and pore volume [16].

Table 1 . The BET surface area, pore diameter and pore volume eggshell powders before and after enzyme immobilisation

	Surface area (m^2 /g)	Pore diameter (nm)	Pore volume (cm^3 /g)
Before immobilization	2.1364	11.8392	6.323×10^{-3}
After immobilization	1.0714	3.7449	1.003×10^{-3}

Physical characterization of eggshell powder before and after RML immobilization was carried out to describe the support capacity or affinity for RML. Figure 2 and 3 showed the N_2 adsorption–desorption isotherm test results to determine the surface area, pore volume and pore diameter of the eggshells. According to IUPAC, both the adsorption isotherms for before and after immobilization were classified to Type II isotherms which were frequently obtained when adsorption takes place on nonporous or macroporous materials. The desorption hysteresis loop occurred in Figure 3 was attributed to the co-existence of intrinsic micropores and inter-crystalline mesoporous structure in eggshell. By comparing Figure 2 and Figure 3, the quantity of nitrogen adsorbed also decreased after lipase immobilization which implies that there was fewer number of pores available after immobilization as most of the pores were occupied by the lipase [17].

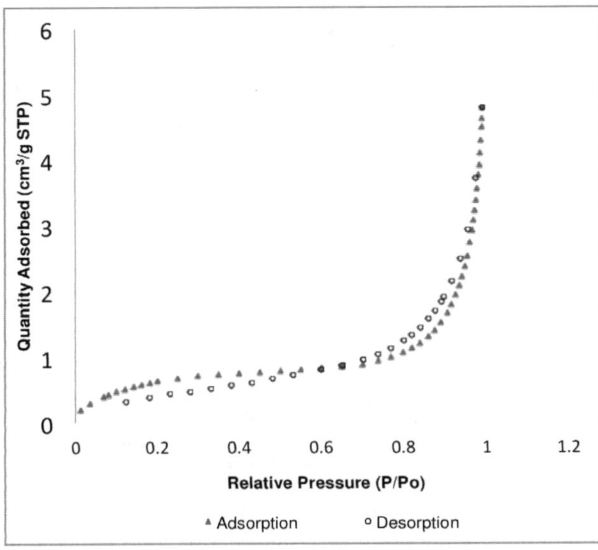

Figure 2. Nitrogen adsorption/desorption isotherms of eggshell powder before

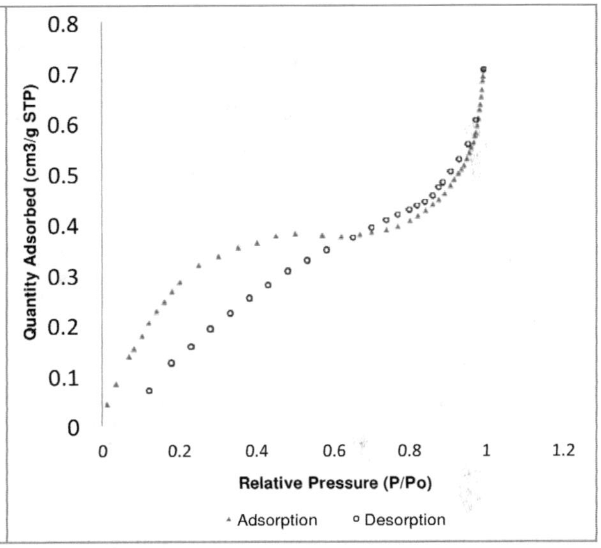

Figure 3. Nitrogen adsorption/desorption isotherms of eggshell powder after immobilization.

Experimental data obtained was fitted into quadratic model and from the subsequent analysis of variance (ANOVA) test, concentration of fructose stearate was presented in term of actual factors as the following:

$$Y = -1.87 + 0.10X_1 - 10.09X_2 + 0.99X_3 + 0.35X_1X_2 - 6.27E^{-3}X_1X_3 - 0.46X_2X_3 - 1.69E^{-3}X_1^2 - 14.37X_2^2 - 0.11X_3^2 \quad \text{Eq. 2}$$

Where X_1 is the temperature (^0C), X_2 is the flow rate (ml/min) and X_3 is the substrate molar ratio of fructose to stearic acid.

Response surface plots using RSM exhibit function of two factors at a time whilst other factors are remained constant. These surface plots are useful in comprehending interaction effects between factors and predicting the yield response for different combination of factor levels. The interaction terms AB (temperature vs flow rate), AC (temperature vs substrate molar ratio of fructose to stearic acid) and

BC (flow rate vs substrate molar ratio of fructose to stearic acid) were observed to have significant effect towards concentration of fructose stearate. For instance, Figure 4 (a) displays the surface plot of an increase in both temperature and flow rate would negatively affect the concentration of fructose stearate. Both the temperature and flow rate have positive effects on the esterification at lowest values in the range of both variables respectively. The temperatures higher than 35^0C was found exceeded the optimal threshold of RML, which might lead to unwinding of immobilized RMLs and progressively inactivated the immobilized RML. Previous study also suggested the breaking of intramolecular hydrogen bonds, van der Waals forces, hydrophobic and ionic interactions that gradually destabilizes and disrupt the RML tertiary structure beyond the optimum temperature, thus probably resulted in a lower product concentration [18]. As shown in the Figure 4 (b), a higher concentration of fructose stearate was reached at a moderate temperature (35 °C) along with highest substrate molar ratio of fructose to stearic acid (3:1). An excess of sugar contributes to an enhanced product yield due to the sugar being a lyoprotectant compound was able to reduce the water activity of the medium, favoring the esterification reaction [19-21]. The optimized conditions (temperature = 37.47^0C, flow rate = 0.74 ml/min and substrate molar ratio of fructose to stearic acid = 2.82:1) were given by the model by using numerical optimization tool with high product concentration of 7.252 x 10^{-1} ± 0.025 mol/L.

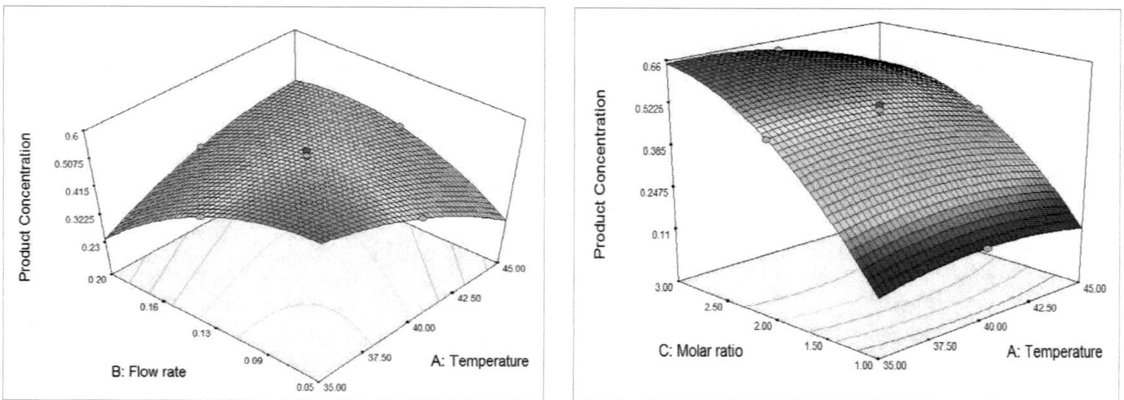

(a) **(b)**

Figure 4 . (a) 3D surface of fructose stearate concentration as a function of temperature (^{0}C) and flow rate (ml/min) at fixed substrate molar ratio of fructose to stearic acid (2:1), **(b)** 3D surface of fructose stearate concentration as a function of temperature (^{0}C) and substrate molar ratio of fructose to stearic acid at fixed flow rate (0.13 ml/min).

4. . Conclusion

A high enzyme immobilization efficiency (63.64%) conducted after the immobilization agreed with the findings from BET, where the adsorption isotherms analysis showed that the Brunauer–Emmett–Teller (BET) surface area, total pore volume and pore diameter of the eggshells decreased significantly after immobilization. Decrement in these surface and porosity values suggested that the lipase was successfully immobilized on the treated eggshells. Subsequently, packed-bed reactor (PBR) was successfully employed in the continuous enzymatic esterification of fructose and stearic acid to produce satisfactory concentration of fructose stearate 7.252 x 10^{-1} mol/L at optimized conditions of 37.47^0C, at flow rate of 0.074 ml/min, using 2.82:1 substrate molar ratio of fructose to stearic acid.

Acknowledgement

This work was supported by bridging research grant Universiti Sains Malaysia (USM) 6316214.

References

[1] Joshi S J and Desai A J 2010 *Biosurfactant's role in bioremediation of NAPL and fermentative production Biosurfactants* (Biosurfactants. Advances in Experimental Medicine and Biology vol 672) ed R Sen (New York: Springer) pp 222-235
[2] Zhang J H, Jiang Y Y, Lin Y, Sun Y F, Zheng S P and Han S Y 2013 *PLoS ONE* **8**(7) e67892
[3] Chattopadhyay S and Sen R 2012 *Process Biochemistry* **47** (5) 749-757

[4] Mateo C, Palomo J M, Fernandez-Lorente G, Guisan J M and Fernandez-Lafuente R 2007 *Enzyme and Microbial technology* **40** (6) 1451-1463
[5] Theron J, Walker J A and Cloete T E 2008 *Crit. Rev. Microbiol.* **34** 43-69
[6] Roger A S 2007 *Adv. Synth. Catal*ysts **349** 1289-1307
[7] Denčić I, Vaan S, Noël T, Meuldijk J, Croon M and Hessel V 2013 *Industrial & Engineering Chemistry Research* **52** (32) 10951-10960
[8] Serri N A, Kamaruddin A and Len K T 2010 *Food and Bioproducts Processing* **88** (2) 327-332
[9] Salleh S, Serri N, Hena S and Tajarudin H 2016 *IOP Conference Series Earth and Environmental Science* **36** (1) 012026
[10] Foo K Y and Hameed B H 2010 *Engineering Journal* **156** 2–10
[11] Gregg S J and Sing K S W 1982 *Adsorption, surface area and porosity* 2nd ed (New York: Academic Press)
[12] Barrett E P, Joyner, L J and Halenda P P 1951 *Journal of the American Chemical Society* **3** (1) 373-380
[13] Al Mijan M, Kim D H and Kwak H S 2014 *Int. J. Food Sci. Technol.* **49** 1751–1757
[14] Rahman M, Netravali A, Tiimob B and Rangari V 2014 *ACS Sustainable Chemistry & Engineering* **2** (10) 2329-2337
[15] Panesar P S, Kumari S and Panesar R 2010 *Enzyme Research* **2010** 473137
[16] Rodrigues D S, Cavalcante G P, Ferreira A L and Gonçalves L R 2008 *Chemical and biochemical engineering quarterly* **22** (1), 125-133
[17] Adnan M, Li K, Xu L and Yan Y 2018 *Catalysts* **8** (3) 96
[18] Tao H, Yang H, Liu X, Ren J, Wang Y and Lu G 2013 *Chemical engineering journal* **225** 686-694
[19] Nawaz M A, Karim A, Bibi Z, Rehman H U, Aman A, Hussain D, Ullah M and Qader S A U 2016 *Journal of the Taiwan Institute of Chemical Engineers* **64** 31-38
[20] Gumel A, Annuar M, Heidelberg T and Chisti Y 2011 *Process Biochemistry* **46** (11) 2079-2090
[21] Soultani S, Engasser J M and Ghoul M 2001 *Journal of Molecular Catalysis B: Enzymatic* **11**(4-6) 725-731.

IOP Conf. Series: Materials Science and Engineering **716** (2020) 012018 doi:10.1088/1757-899X/716/1/012018

Determination of nucleation assay for anti-urolithiasis activity from bagasse *Musa acuminate x balbisiana Colla* cv. Pisang Awak Legor methanolic extracts using uv-spectrometer and size measurement.

M A Zarin[ab]*, J S Tan[a]*, R Ahmad[a], N Z Jin[a], N F H A Aziz[a]

[a]Division of Bioprocess Technology, School of Industrial Technology, Universiti Sains Malaysia, 11800 Minden, Pulau Pinang, Malaysia.
[b] Laboratory of Vaccines and Immunotherapeuthics, Institute of Bioscience, Universiti Putra Malaysia, 43400 UPM Serdang, Selangor, Malaysia.

*Email: jooshun@usm.my; mazniaz@upm.edu.my

ABSTRACT

Determination of inhibition of calcium oxalate (CaOx) crystallisation *in vitro* by nucleation assay is based on the rates of nucleation by comparing the slope of the turbidity of a control system with that of one exposed to the extract, which was found to be inaccurate when the CaOx crystals were formed and precipitated. A need exists for improved methods to determine the inhibition activity in nucleation assay for the study of anti-urolithiasis activity. In this study, the size reduction of CaOx crystals after treated with methanolic extract of *Musa acuminate* x *balbisiana Colla* cv. Pisang Awak Legor bagasse was measured under light microscopy and correlated to the turbidity and percentage of inhibition. Different concentrations of bagasse extract (2, 10, 30, 50 and 100 mg/ml) were investigated and the result shows that high concentration of extract would promote reduction in CaOx's stone size, but it increased the amount of sediment crystals, resulting high turbidity. The stone size was reduced up to 93.76±0.19% in the presence of 100 mg/mL extract while the inhibition percentage was negative (-102.17±0.04%) as the optical density (OD) of the extract was higher than control in the turbidity calculation. The results of this study are expected to provide an understanding on the way of calculating the activity in nucleation assay.

1. INTRODUCTION

The remedies have been used for a long time in treating kidney stone disease. Most of the remedies used in traditional medicine systems are taken from plants although they do not have strong systematics clinical evidence yet have a positive effect to the patients except for some composite plants and herbal drugs (Yadav et al. 2011) such as *Herniaria hirsuta* where the extract can decrease the crystal size (Atmani and Khan 2000), *Bergenia ligulata* (Aziz et at. 2005), *Piper nigrum* (Manish et al. 2011), *Dolichos biflorus* (Garimella et al. 2001), *Plantago major* (Verma et al. 2014) etc. Most of the previous findings were done *in vivo* either in animal or human (Yadav et al. 2011).

Banana was a fruit largely consumed all over the world. Different parts of banana plant (*Musa*) including pseudo-stem were known to be used in Ayurveda and other traditional folklore medicine for treatment of various diseases including kidney stone (Kalpana et al. 2013). Studies using banana pseudo-stem has not yet explored extensively in clinical system especially from bagasse from *Musa acuminate x balbisiana Colla* cv. Pokok Awak Legor. This type of *Musa* was in ABB Group.

Nucleation assay is one of the most important steps in stone formation. Generally, the percentage of inhibition stone was calculated using the equation established previously by Hennequin et al. (1993), where the value was obtained from optical density measurement at 620nm using spectrometer. Unfortunately, the method was not suitable when increasing the concentration of extract as the Caox formation become small particles and increased the turbidity of solution more than control without treatment. It is impossible to mimic all the processes involved in stone formation *in vitro*, thus it is important to choose the most appropriate approach for measuring the inhibition process for *in vitro* study. Hence, this study was conducted to prove that different assessment needs to implement to measure the nucleation assay activity and obtained accurate results.

2. MATERIALS AND METHODS

2.1 Plant Sample Collection and preparation of extract

Large quantities of the pseudo-stem *Musa acuminate x balbisiana Colla* cv. Pisang Awak Legor have been collected from the village in Padang Serai, Kedah, Malaysia. Each pseudo-stem was cleaned with deionized water and pressed using sugarcane extractor to remove the juice. The bagasse was frozen prior to freeze drying and grinding. 5 g of bagasse was soaked in 300 mL of 80% (v/v) methanol at room temperature for a while and filtered through Whatman filter paper No. 1. The extraction was repeated several times until colorless. The extract was evaporated in rotary evaporator at a temperature not exceeding 40°C prior to freeze dry.

2.2 Nucleation Assay

Nucleation assay was done according to Rajeswari et al. (2013) method with some modification. Calcium chloride (5 mmol/L) and sodium oxalate (7.5 mmol/L) solution were prepared in a buffer containing Tris 0.05 mol/L and NaCl 0.15 mol/L at pH 6.5 and 37^{0}C. A total of 95 µl calcium chloride solutions were added to the different concentration of extract (2, 10, 30, 50, 100 mg/ml) in 96 wells plate. Crystallization was started by adding 95 µl of sodium oxalate. The absorbance (optical density, OD) was recorded at 620 nm after 6 hours of incubation. The size of the CaOx stone was observed and measured using microscope for comparison.

IOP Conf. Series: Materials Science and Engineering **716** (2020) 012018 doi:10.1088/1757-899X/716/1/012018

2.3 Microscopic observation

The CaOx crystals formed with or without the extracts were observed by using light microscopy (Olympus CX41) equipped with software analysis LS starter and camera Olympus U-CMAD3 (Japan) at 40X magnification.

2.4 Statistical Analysis

The data obtained was analyzed by two-ways using GraphPad Prism 7. A value of $p < 0.05$ was considered significant in all cases.

3. RESULTS AND DISCUSSION

The methanol was selected as a solvent for extraction based on a previous study by Chilivery et al. (2016) where it showed that methanolic extract from *Argemone mexicana* L. leaves had highest anti-urolithiatic activity compared to petroleum ether, chloroform and water. Moreover, methanol has a polarity index of 5.1 and good in extracting both polar and non-polar substances and easier to concentrate the extract using rotavapor as the boiling point is just 65^0C. The relationship between OD and size of CaOx against extract concentration was shown in Table 1. Increase in extract concentrations from 2 mg/mL to 100 mg/mL increased the turbidity, in terms of OD value from 0.167 to 0.744 while the size of Caox crystal was decreased from 24.9 µm to 3.13 µm. On the other hand, the control (without extract) has higher turbidity and larger CaOx size. The assessment method in anti-urolithiasis activity by measuring the optical density using UV spectrometer in nucleation assay in vitro was established by Hennequin et al. (1993). This method has been followed by most of the studies (Rajeswari et al. 2013; Saha and Verma 2013; Atmani and Khan 2001). The method has been slightly modified and the equation of inhibition percentage used was $[(Ac - As)/Ac)]*100$ where, Ac is absorbance without treatment and As is absorbance with treatment. However, our result was contrary to this method, as the OD was not linearly correlated to the size of CaOx (Figure 1). Figure 1 showed that OD increased linearly with the increase in extract concentration, while the size of CaOx was reduced markedly from 2 mg/mL to 10 mg/mL of extract but slowed down afterwards. The result shows that high concentration of extract caused the formation of CaOx's stones into smaller crystals but higher amount, resulting higher turbidity. Hence, the inhibition percentage of nucleation assay using the equation established in by Hennequin et al. (1993) was not suitable in this study. Two-ways ANOVA analyses show the interaction of turbidity and size with concentration of extract is significantly different with p value < 0.0001.

Table 1: The optical density value and size of CaOx stone at 6 hours of incubation

Sample (mg/mL)	Turbidity (OD)	Size (µm)
Control (0)	0.368±0.04	50.10±0.53
2	0.167±0.03	24.9±1.13
10	0.280±0.01	7.70±0.40
30	0.348±0.02	6.02±0.32
50	0.471±0.03	4.12±0.54
100	0.744±0.04	3.13±0.19

IOP Conf. Series: Materials Science and Engineering **716** (2020) 012018 doi:10.1088/1757-899X/716/1/012018

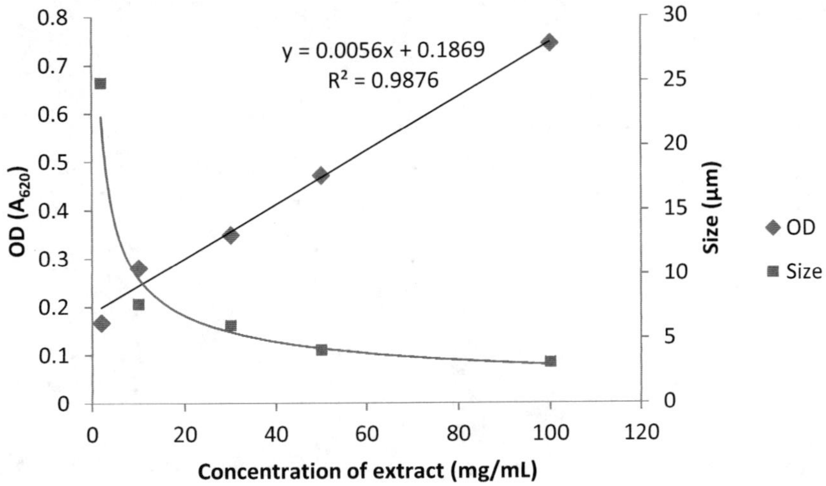

Figure 1: Relation between absorbance at 620nm with CaOx size (µm) at difference concentration of extract incubated at 37^0C for 6 hours.

Figure 2 shows the CaOx crystals size under microscopic observation at 40X magnification. The size of stone formation was measured and calculated the inhibition percentage using the equation [(Sc – Ss)/Sc)]*100 where, Sc is size of CaOx without treatment and Ss is size of CaOx with treatment. The size of CaOx stone formation at different concentration of extracts was decreased from 56.10±1.84 µm to 2.37±0.49 µm at 0 to 100 mg/mL, respectively. The results were compared to the nucleation assay using OD method as shown in Table 2. The inhibition percentage by size reduction method was 93.76±0.19% in the presence of 100 mg/mL extract while the inhibition percentage became negative after treatments with extract concentration exceed 30 mg/mL (-27.99% at 50 mg/mL and -102.17±0.04% at 100 mg//mL) in OD method. This is due to the interference of the crystals towards the detection of spectrophotometer which resulted the OD was higher than control in the turbidity calculation.

IOP Conf. Series: Materials Science and Engineering **716** (2020) 012018 doi:10.1088/1757-899X/716/1/012018

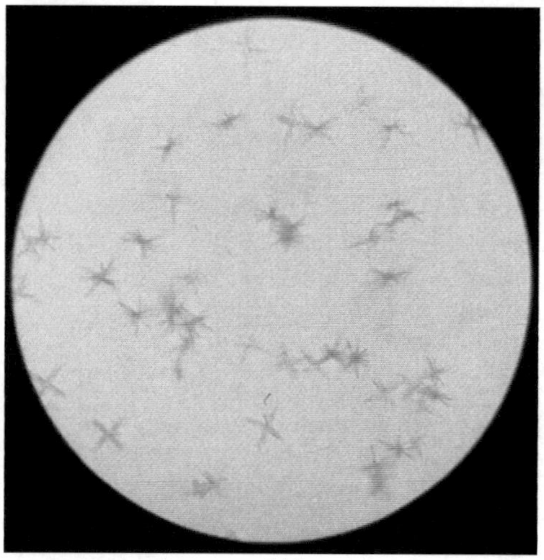

(A)

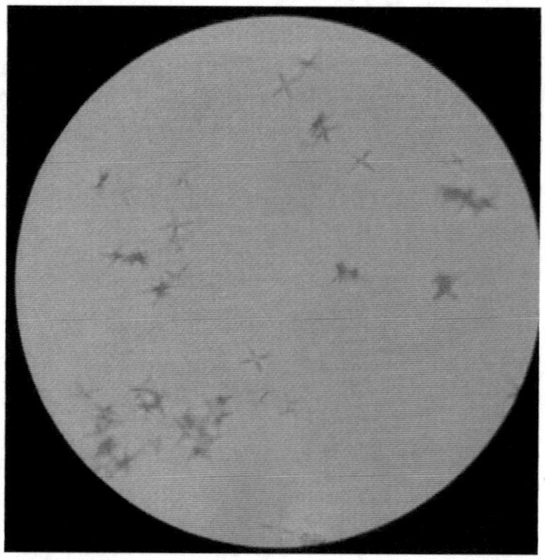

(B)

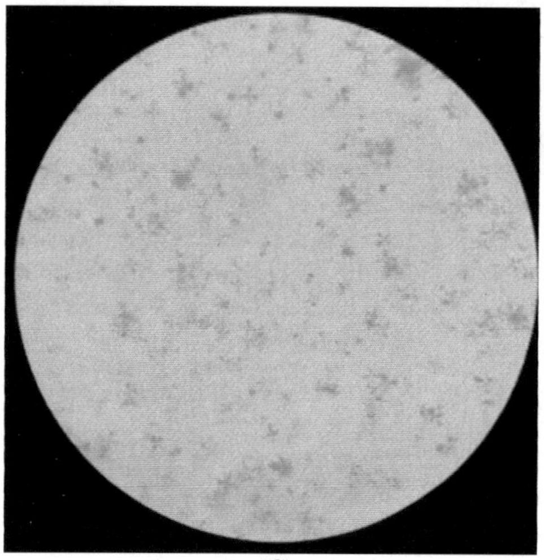

(C)

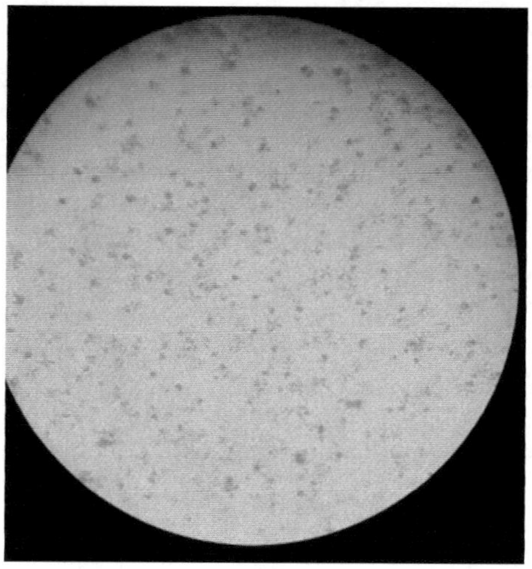

(D)

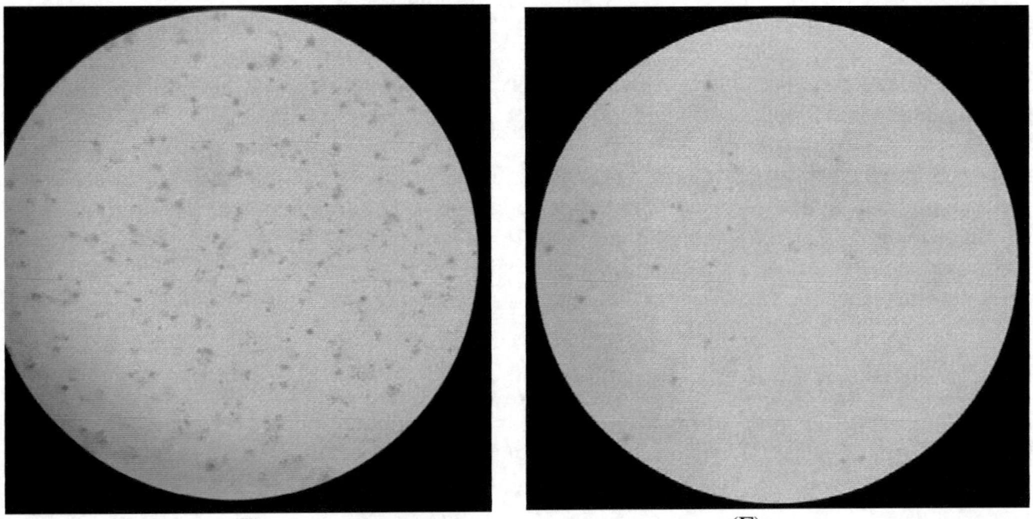

(E) (F)

Figure 2: Stone formation under light microscopy at 40X magnificant in nucleation assay at different concentration of bagasse *Musa acuminate x balbisiana Colla* cv. Pisang Awak Legor methanolic extract incubated at 37^0C for 6 hours. (A) 0 mg/mL (B) 2 mg/mL (C) 10 mg/mL (D) 30 mg/mL (E) 50 mg/mL (F) 100 mg/mL

Table 2: Percentage of inhibition and size reduction at different concentration of extract

Concentration	Inhibition by OD (%)	Inhibition by size (%)
2	54.62±0.04	50.30±1.13
10	23.91±0.01	84.63±0.4
30	5.43±0.02	87.98±0.3
50	-27.99±0.03	91.78±0.54
100	-102.17±0.04	93.76±0.19

The measurement of the CaOx stone size is proposed in this study as the OD value keeps on increasing with the increase of the extract concentration. However, the size became smaller causes the inability to comply with the commonly used equation in determining the percentage of inhibition. In this case, the increasing of the turbidity were due to the small particles of the CaOx stone after treated with the extracts and resultant the reduction of inhibition percentage. Previous studied by Pachana et al. (2010) showed that the higher concentration of *Tribulus terrestris* extract inhibits the formation of COM (calcium oxalate monohydrate), and promotes the formation of COD (calcium oxalate dehydrate) in the treated group compared to control. This could be one of the reasons that the OD value was increased with the increment of extract concentration, resulted in the negative inhibition percentage. Atmani and Khan (2001) also found that *Herniaria hirsuta* extract promoted the nucleation of CaOx crystals, increasing their number but decreasing their size based on the scanning electron microscopy (SEM) urine analysis which could be another reason the absorbance value increasing when increased the concentration of extract as the crystals formed were mostly COD. The finding also showed that size reduction gave an advantageous as it can preventing urinary stone formation by inducing the excretion of small particles the kidney and reducing the chance of retention in the urinary tract. Thus, this new approach in assessment of stone inhibition was applied to obtain the accurate result.

Theoretically, stones size less than 5 mm have a high possibility of spontaneous passage through the urinary tract and out of the body on their own with little or no pain which can take up to 40 days or the patient might undergo surgery to remove the stone (Coll et al. 2002). Measurement of Caox stone size in *in vitro* study will be easier in estimation or screening the potential selected plant extracts as anti-urolithiasis agents since the researcher could determine the accurate size reduction.

CONCLUSION

In in vitro study, determination of nucleation assay would be more suitable by using size determination compared to the absorbance (OD value) due to the interference of the crystals. The accuracy of the activity especially in higher plant extracts concentration could be increased using size determination.

REFERENCES

[1] Atmani F and Khan SR 2001 Effect of an extract from *Herniaria hirsute* on calcium oxalate crystallization in vitro Braz J Urol Int. 85: 621-625.
[2] Aziz SA, See TL, Khuay LY, Osman K, Abu Bakar MA 2005 *In vitro* Effects of *Plantago major* Extract on Urolithiasis. Malays J Med Sci. 12(2): 22-26.
[3] Coll DM, Varanelli MJ, Smith RC 2002 Relationship of spontaneous passage of ureteral calculi to stone size and location as revealed by unenhanced helical CT. AJR Am. J. Roentgenol. 178:101–103.
[4] Chilivery RK, Alagar S, Darsini TP 2016. In Vitro Anti-Urolithiasis Potentials of *Argemone mexicana* L. Leaves. Curr Clin Pharmacol. 11(4): 286-290.
[5] Garimella TS, Jolly CI, Narayanan S 2011 *In vitro* Studies on Antilithiatic Activity of Seeds of *Dolichos biflorus Linn* and Rhizomes *of Bergenia ligulata* Wall. Phytother Res. 15(4): 351-355.
[6] Manish AP, Paras KP, Avinash KS 2011 Inhibition of Calcium Oxalate Crystallization by the fruit extracts of *Piper nigrum* L. Pharmacologyonline. 2: 1169-1177.
[7] Pachana K, Wattanakornsire A, Nanuam J 2010 Application of small caltrops (*Tribulus terrestris*) to inhibit calcium oxalate monohydrate crystallization. Sci Asia. 36: 165-168.
[8] Rajeswari P, Rajeswari G, Jabbrirulla SK, Vardhan IV 2013 Evaluation of *in vitro* anti-urolithiasis activity of *Convolvulus arvensis.* Int J Pharm Pharm Sci. 5(3): 599-601.
[9] Saha S, Verma RJ 2013 Inhibition of calcium oxalate crystallization in vitro by an extract of *Bergenia ciliate*. Arab J Urol. 11:187–192.
[10] Verma P, Gauttam V, Kalia AN 2014 Comparative Pharmacognosy of Pashanbhed. J Ayurveda. Integr Med. 5(2):104-108.
[11] Yadav RD, Jain SK, Alok S, Mahor A, Bharti JP, Jaiswal M 2011 Herbal Plants Used in the Treatment of Urolithiasia: A Review. Int J Pharm Sci Res. 2(6): 1412-1420.

ACKNOWLEDGEMENT

The Authors would like to thank to Universiti Putra Malaysia for scholarship fund to Mazni Abu Zarin. Ng Zhang Jin is recipient of USM Fellowship RU(1001/CIPS/AUPE001) from Universiti Sains Malaysia.

IOP Conf. Series: Materials Science and Engineering **716** (2020) 012019 doi:10.1088/1757-899X/716/1/012019

Response surface optimization of the forward extraction of jacalin from jackfruit seeds using AOT/isooctane reverse micellar system

S F S Mohamad[1], F M Said[1], M S A Munaim[1], S Mohamad[1] and W M A W Sulaiman[2]

[1]Faculty of Chemical & Process Engineering Technology, Universiti Malaysia Pahang, Lebuhraya Tun Razak, 26300 Kuantan, Pahang, Malaysia.
[2]Department of Basic Medical Sciences, International Islamic University Malaysia, Jalan Sultan Ahmad Shah, 25200 Kuantan, Pahang, Malaysia.

Email: fathiyah@ump.edu.my

Abstract. Jacalin is the major protein contained in the crude extract of jackfruit (*Artocarpus heterophyllus*) seed that specifically recognizes and binds reversibly to galactose. Conventionally, purification of jacalin is carried out using the tedious and costly chromatographic techniques. In this study, extraction of jacalin from jackfruit seed crude extract were done using the sodium bis(2-ethylhexyl) sulfosuccinate (AOT)-based reverse micellar system. Reverse micellar extraction is an attractive alternative for downstream processing of various proteins. A successful reverse micellar extraction consists of two basic steps: forward and backward extraction. Forward extraction transfers a target protein from an aqueous solution into the reverse micellar solution, while backward extraction releases the protein from the reverse micelles structure into a new aqueous solution. The effects of the aqueous phase pH, NaCl concentration and AOT concentration on the forward extraction efficiency (FEE) are investigated using the response surface methodology (Box-Behnken Design). The main effects and interactions of the parameters are analyzed through the 3D surface plots. The optimum conditions for forward extraction were determined as follows: aqueous phase pH 4.58, 125 mM NaCl and 40 mM AOT. Under the optimal conditions, the FEE reached 88.04±1.30%, closer to 87.99% predicted by the model. The results indicated that AOT/isooctane reverse micelle system is effective in extracting jacalin from the jackfruit seed crude extract and verified the practicability of the BBD model for optimizing the main parameters in the forward extraction of jacalin.

Keywords: Optimization; Reverse micelles; Forward extraction; Jacalin; sodium bis(2-ethylhexyl) sulfosuccinate (AOT)

1. Introduction

Jacalin is a tetrameric two-chain lectin with the molecular weight of 65-66 kDa isolated from the jackfruit seed *(Artocarpus heterophyllus)* [1]. Jacalin has been widely investigated due its numerous biological properties, such as inhibition and binding with specific regions of HIV [2], ability to recognize particular immunoglobulin A1 from human serum [3], stimulation of different T cell functions [4,5] and specific binding with tumor associated Thomsen-Friedenreich (TF) antigen [6]. Conventional approaches to purify jacalin from the jackfruit seeds involved various chromatographic techniques, i.e. affinity chromatography on immobilized IgA-Sepharose column [4], Minileak-melibiose [3] or cross-linked guar gum matrix [7], ion-exchange chromatography [8] and partition chromatography on analytical C4 reversed-phase high performance liquid chromatography [9]. The chromatographic techniques are mostly preferred due to the ability to produce highly purified jacalin, however the methods are laborious, time-consuming and difficult to scale-up for industrial application.

Reverse micelles are thermodynamically stable, spontaneous aggregates formed when amphiphilic surfactant molecules were dissolved in organic solvent [10]. Their inner core contains nanometer-sized water pools that can solubilize biomolecules such as enzymes and proteins [11,12]. Extraction by reverse micelles or reverse micellar extraction (RME) has been one of the most reliable alternative method for downstream processing of various types of proteins. Compared to the conventional purification method of jacalin, RME has many attractive features, including large interfacial area, less energy requirement, continuous mode of operation, low-cost and easy to scale-up [10]. A reverse micellar extraction can be conducted by phase transfer method consisting of two basic steps, forward and backward extraction. Forward extraction transferred a target protein contained in an aqueous solution into a reverse micellar solution while backward extraction released the protein from the reverse micelles structure into a new aqueous solution [13].

The efficiency of the reverse micellar system to selectively transfer target protein from its mixture was controlled by several parameters, such as concentration and type of surfactant, aqueous phase pH, concentration and type of salt, water content, size of reverse micelles, temperature and many more [14,15]. Hence, optimization of the process parameters is critical in maximizing the extraction efficiency of the target protein in reverse micellar system. Response surface methodology (RSM) is a collection of statistical methods based on multivariate nonlinear model that has been extensively applied in various fields of research to determine the effects of several variables and to find the optimum process conditions for the selected process [15,16]. RSM has been applied recently to optimize variables affecting the extraction efficiency of *Phaseolus vulgaris* lectin in AOT/isooctane reverse micellar system [17]. However, until now, there has been no such report focusing on the screening and optimization of the process parameters affecting the forward extraction of jacalin from the crude extract of jackfruit seeds.

Therefore, this study was aimed to maximize the forward extraction efficiency and optimize the forward extraction conditions of jacalin isolated from the jackfruit seed crude extract by response surface methodology (RSM). To do that, a Box-Behnken design (BBD) is implemented and a second order model is developed to predict and optimize the desired response.

2. Methods

2.1. *Sample collection and preparation of jackfruit seed crude extract*

Jackfruits (*Artocarpus heterophyllus* Lam.) used in this study were of local variety known as Mastura, bought from a fruit market in Kuantan, Pahang, Malaysia. Jackfruit seeds were manually separated from the ripe flesh, washed, and thinly sliced before dried in an oven at 37°C for about 24 h. The dried seeds were ground into powder using a mixer-grinder and passed through an 80-mesh sieve. The jackfruit seed crude extract was prepared by mixing the fine jackfruit seed powder with 10 mM phosphate-buffered saline (PBS) containing 0.15 M NaCl (1:10 w/v, pH 7.4) and gently stirred at 4°C for 24 h. After 20 min centrifugation at 10000 rpm, the clear supernatant was collected, filtered through a 0.45 μm membrane and used in the next step as the crude extract.

2.2. *Forward extraction procedure*

Forward extraction was carried out using the phase transfer method in screw-capped centrifuge tubes. In this study, 5 ml of organic (AOT dissolved in isooctane) and 5 ml of aqueous solutions (dilutions of crude extract in buffer solution) was slowly mixed with a magnetic stirrer at 300 rpm for 20 min in room temperature. After mixing, the solution was centrifuged at 5000 rpm for 10 min to achieve a clear phase separation between two phases. Both the organic and aqueous phases were collected and the protein concentrations in each phase were measured.

2.3. *Optimization of forward extraction by RSM*

A response surface methodology (RSM) based on a Box-Behnken design (BBD) was applied to study the optimum conditions for forward extraction of jacalin using AOT reverse micellar system. Based on an one factor at a time (OFAT) study and a factorial analysis [18,19], the aqueous phase pH (pH 4.2 – pH 5.2), AOT concentration (10 – 40 mM) and NaCl concentration (100 – 150 mM) were examined in three different levels (-1, 0 and +1) as shown in Table 1.

Table 1. Independent variables and their respective levels.

Independent variables	Symbols	Coded Factor Levels		
		−1	0	+1
pH of the aqueous phase	A	4.2	4.7	5.2
AOT concentration (mM)	B	10	25	40
NaCl concentration (mM)	C	100	125	150

A total of 17 experiments including 5 repetitions of the center points were conducted in triplicates. All experimental results expressed as mean ± standard deviation was used in the data analysis. The experimental design matrix, and data analysis for optimization procedure were performed using the obtained using the Design Expert Version 10 software® (Version 10.0.1.0 64-bit Stat-Ease Inc., Minneapolis, USA). Mean values were considered significantly different when $p<0.05$. The effects of these independent variables on the response, Y can be predicted using a polynomial regression model of second order (Equation 1) as follows:

$$Y = \beta_0 + \sum_{i=1}^{3} \beta_i X_i + \sum_{i=1}^{3} \beta_{ii} X_i^2 + \sum_{i<i=1}^{3} \beta_{ii} X_i X_j \tag{1}$$

where Y is the response calculated by the model, β_0, β_i, β_{ii} and β_{ij} are the coefficients of intercept, linear, quadratic and interaction terms, respectively, while X_i and X_j indicated the independent variables.

2.4. *Analysis of jacalin concentration*

Jacalin concentration in both feed and aqueous phase after forward extraction were quantified using Lowry assay with bovine serum albumin (BSA) as a standard [20]. Mass balance was used to estimate the amount of jacalin in the organic phase. The samples were analyzed against respective blanks performed with an aqueous phase voided of crude extract. Jacalin concentrations were expressed in mg/ml. Average values of duplicate readings of jacalin concentrations were used to estimate the efficiency of forward extraction (FEE) using the following Equation (2).

$$FEE(\%) = \frac{C_{org}}{C_0} \times 100 \tag{2}$$

where C_{org} is the concentration of protein in the organic phase after forward extraction and C_0 is the initial concentration of protein in the aqueous phase before forward extraction was performed.

3. Results and Discussion

3.1. Statistical analysis and model fitting

The effects of three forward extraction parameters: aqueous phase pH (A), AOT concentration (B) and NaCl concentration (C) on the forward extraction efficiency (FEE) of jacalin were studied using a Box-Behnken design of experiment. Table 2 shows the experimental matrix design of the total 17 runs, along with the experimental and predicted results obtained.

Table 2. Design matrix and experimental responses of the BBD.

Run No.	Independent parameters			Response (Y)	
	Actual values			Forward extraction efficiency (%)	
	A	*B*	*C*	Experimental (Y_e)	Predicted (Y_p)
1	5.2	25	100	44.44 ± 0.16	43.79
2	5.2	25	150	41.67 ± 0.22	44.12
3	4.2	40	125	75.41 ± 0.24	76.14
4	4.7	25	125	83.03 ± 0.11	81.76
5	4.2	25	100	73.14 ± 0.12	70.70
6	4.2	25	150	70.44 ± 0.22	71.09
7	4.7	25	125	84.16 ± 0.23	81.76
8	5.2	40	125	57.46 ± 0.11	56.39
9	4.7	40	100	81.77 ± 0.91	83.49
10	4.7	25	125	82.24 ± 2.28	81.76
11	4.7	10	100	74.65 ± 0.23	76.03
12	4.7	25	125	80.21 ± 0.34	81.76
13	4.7	40	150	83.89 ± 0.11	82.51
14	4.7	10	150	79.44 ± 0.34	77.73
15	4.7	25	125	79.18 ± 0.80	81.76
16	5.2	10	125	43.81 ± 0.35	43.08
17	4.2	10	125	76.15 ± 0.22	77.22

The experimental results (FEE) obtained from the BBD were analyzed and fitted to the following second-order polynomial equation (Equation 3):

$$Y = 81.76 - 13.47A + 3.06B + 0.18C + 3.60AB - 0.014AC - 0.67BC - 20.54A^2 + 1.98B^2 - 3.80C^2 \tag{3}$$

In this study, the parameters influencing the response of forward extraction of jacalin are in the following order: aqueous phase pH > AOT concentration > NaCl concentration. Furthermore, the lack of fit value of 0.2471 implied that the lack of fit was not significantly correlated with the pure error, hence further validates the model adequacy for predicting the forward extraction efficiency of jacalin.

The coefficient of variation (CV) of the model, defined as the ratio of the standard error of estimate to the mean value of observed response, was calculated as 3.46 %. Generally, a model can be considered reasonably reproducible if its CV is not greater than 10 %. The Adeq. Precision of the test calculated (21.40) was higher than the desired value (4.00), indicating low signal-to-noise ratio and the model can be used to navigate design space. These results proved that the developed polynomial model could adequately describe the real relationship among the parameters chosen and therefore, can be used for the subsequent process optimization.

3.2. Effect of parameters and the interaction between parameters on FEE of jacalin

Figure 1(a) shows predicted values of FEE by varying pH of the aqueous phase (A) and AOT concentration (B), while fixing the NaCl concentration (C) at 125 mM. At all levels of AOT concentration, a significant quadratic effect of pH was observed. Initially, the FEE was found to increase with the increasing pH of the aqueous phase pH until the FEE reached maximum value at pH 4.7. However, further increasing of the aqueous phase pH beyond pH 4.7 resulted in the decreasing of the FEE. Similar results of increase in extraction efficiency with the increase in pH were also observed by He et al. [17] while conducting optimization of forward extraction of lectin from Phaseolus vulgaris [17] using the similar AOT reverse micellar system. This phenomenon may be due to the strong electrostatic interaction in AOT/isooctane reverse micelles, which only solubilizes the target protein when the pH of the aqueous phase was below the isoelectric point (pI) of the target protein [21]. According to He et al. [22], pH values above the pI of the proteins might lead to a decrease in charge density and further yield loss of the proteins.

Figure 1(b) demonstrated that the both aqueous phase pH (A) and NaCl concentration (C) had significant quadratic effects on the FEE, while the AOT concentration (B) was fixed at 25 mM. Overall, at all levels of NaCl concentration, the FEE observed at the lowest pH value (pH 4.2) was higher than the FEE observed at the highest pH value (pH 5.2). Initially, increasing of aqueous phase pH value from pH 4.2 until pH 4.7, resulted in slight increase of FEE until a maximum FEE was found at 125 mM NaCl. These results suggested that the favorable interaction between jacalin and surfactant head groups was enhanced through the addition of NaCl, known as the salting-in effect. However, an opposite trend could be observed with further increase of the NaCl concentration and pH up to the maximum value at pH 5.2. It may have occurred due to the reduction of electrostatic repulsion between charged head groups of the AOT surfactants in high NaCl concentration, thereby decreasing the size of reverse micelles. Consequently, larger protein molecules might be excluded at a higher NaCl concentration, due to the Debye length reduction and size exclusion effect [22].

Figure 1(c) shows the interactive effects of the AOT concentration (B) and the NaCl concentration (C) on the FEE. At all levels of NaCl concentration, the FEE was found to increase with an increasing of AOT concentration Similar trend of increasing extraction efficiency with increasing surfactant concentration was also observed in extraction and purification of 7s globulin subunits from soybean protein [23], probably due to the increase number of reverse micelles and the corresponding size of reverse micelles, as there were more water cores in which jacalin could be solubilized [24].

IOP Conf. Series: Materials Science and Engineering **716** (2020) 012019 doi:10.1088/1757-899X/716/1/012019

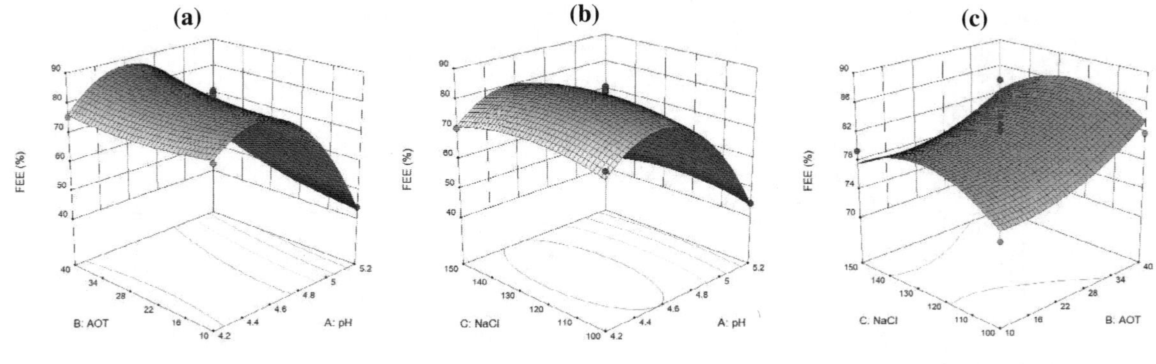

Figure 1. Response surface curves showing the effects of aqueous phase pH (A), AOT concentration (B) and NaCl concentration (C) on the forward extraction efficiency of jacalin.

3.3. Verification of optimum condition

The validation of the regression model equation for predicting the optimum response values was tested by conducting three confirmation experiments using the recommended optimum conditions obtained from the numerical RSM optimization approach. The optimum process conditions for aqueous phase pH (A), AOT concentration (B) and NaCl concentration (C) were pH 4.58, 40 mM and 125 mM, respectively (Table 3). Under these optimum process conditions, the experimental response (%FEE) obtained was 88.04±1.30%, which is closely related to the predicted FEE of 87.99%.

Table 3. Verification of optimum forward extraction condition.

Parameter	Optimum process conditions		Predicated response (% FEE)	Actual response (% FEE)	Deviation (%)
	Coded	Uncoded			
Aqueous phase pH	-0.241	pH 4.58	87.99	88.04 ± 1.30	0.06
AOT concentration	1.000	40 mM			
NaCl concentration	0.024	125 mM			

4. Conclusion

BBD-based RSM was successfully used to optimize the conditions of forward extraction of jacalin using AOT reverse micelles. The developed model for FEE exhibited non-significant lack of fit and R2 value of 0.9879. The surface graphs indicated that maximum forward extraction efficiency could be obtained by extracting jacalin with aqueous phase pH of 4.58, containing 125 mM NaCl and AOT concentration of 40 mM. Under these optimum conditions, the forward extraction efficiency observed was 88.04 ± 1.30%, slightly higher than the predicted value of 87.99%. From these experimental results, it can be concluded that extraction using AOT reverse micelles were effective to extract jacalin from the crude extract of jackfruit seed powder and that the regression model equation was suitable and adequate for predicting the optimum response values in the forward extraction process.

References

[1] Kabir S 1998 Jacalin: A jackfruit (*Artocarpus heterophyllus*) seed-derived lectin of versatile applications in immunobiological research *J. Immunol. Methods* **212** 193–211

[2] Corbeau P, Pasquali J-L and Devaux C 1995 Jacalin, a lectin interacting with O-linked sugars and mediating protection of $CD4^+$ cells against HIV-1, binds to the external envelope

glycoprotein gp120. *Immunol. Lett.* **47** 141–143
[3] Hagiwara K, Collet-Cassart D, Kunihiko K and Vaerman J-P 1988 Jacalin: Isolation, Characterization, and Influence of Various Factors on Its Interaction with Human IgA1, As Assessed by Precipitation and Latex Agglutination. *Mol. Immunol.* **25** 69–83
[4] Bunn-Moreno MM and Campos-Neto A 1981 Lectin(s) extracted from seeds of *Artocarpus integrifolia* (jackfruit): potent and selective stimulator(s) of distinct human T and B cell functions. *J. Immunol.* **127** 427–429
[5] Lafont V, Hivroz C, Carayon P, Dornand J and Favero J 1997 The lectin jacalin specifically triggers cell signaling in $CD4^+$ T lymphocytes. *Cell Immunol.* **29** 23–29
[6] Mahanta SK, Sanker S, Rao N V, Swamy MJ and Surolia A 1992 Primary structure of a Thomsen-Friedenreich-antigen-specific lectin, jacalin [*Artocarpus integrifolia* (jackfruit) agglutinin]. Evidence for the presence of an internal repeat. *Biochem. J.* **284** 95–101
[7] Roy I, Sardar M and Gupta MN 2005 Cross-linked alginate-guar gum beads as fluidized bed affinity media for purification of jacalin. *Biochem. Eng. J.* **23** 193–198
[8] Ruffet E, Paquet N, Frutiger S, Hughes GJ and Jaton J-C 1992 Structural and electron-microscopic studies of jacalin from jackfruit (*Artocarpus integrifolia*) show that this lectin is a 65 kDa tetramer. *Biochem. J.* **286** 131–134
[9] Young NM, Johnston RAZ and Watson DC 1991 The amino acid sequences of jacalin and the *Maclura pomifera* agglutinin. *FEBS Lett.* **282** 382–384
[10] Gangadharappa BS, Dammalli M, Rajashekarappa S, Pandurangappa KMT and Siddaiah GB 2017 Reverse micelles as a bioseparation tool for enzymes. *J. Proteins Proteomics* **8** 105–120
[11] Murugesan S, Iyyaswami R, Kumar SV, Surendran A and Vijay S 2017 Anionic surfactant based reverse micellar extraction of l-asparaginase synthesized by *Azotobacter vinelandii*. *Bioprocess Biosyst. Eng.* **40** 1163–1171
[12] Hemavathi AB, Hebbar HU and Raghavarao KSMS 2012 Reverse micellar extraction of bioactive compounds for food products. In: Libovka N, Vorobiev E, Chemat F (eds) *Enhancing Extraction Processing Food Industries* CRC Press, pp 399–436
[13] Leser ME, Wei G, Luisi PL and Maestro M 1986 Application of reverse miclles for the extraction of proteins. *Biochem. Biophys. Res. Commun.* **135** 629–635
[14] Li S and Cao X 2014 Extraction of tea polysaccharides (TPS) using anionic reverse micellar system. *Sep. Purif. Technol.* **122** 306–314
[15] Prabhu AA, Chityala S, Garg Y and Dasu VV 2017 Reverse micellar extraction of papain with cationic detergent based system: An optimization approach. *Prep. Biochem. Biotechnol.* **47** 236-244
[16] Chuo SC, Ahmad A, Mohd-Setapar SH, Mohamed SF and Rafatullah M 2018 Utilization of green sophorolipids biosurfactant in reverse micelle extraction of antibiotics: Kinetic and mass transfer studies. *J Mol. Liq.* **276** 225–232
[17] He S, Shi J, Walid E, Zhang H, Ma Y and Xue SJ 2015 Reverse micellar extraction of lectin from black turtle bean (*Phaseolus vulgaris*): Optimisation of extraction conditions by response surface methodology. *Food Chem.* **166** 93–100
[18] Sy Mohamad SF, Mohd Said F, Abdul Munaim MS and Mohamad S 2019 Effect of key parameters on jacalin extraction from aqueous phase into anionic reverse micelles. *Indones. J. Chem.* **19(3)** 737–746
[19] Sy Mohamad SF, Mohd Said F, Abdul Munaim MS, Mohamad S and Wan Sulaiman WMA 2018 Fractional factorial analysis for identifying significant factors in extraction of jacalin from jackfruit seeds using anionic reverse micellar system. *Int. Conf. in Chemical Science and Engineering (ICCSE)* 7-8 Nov. Melaka, Malaysia.
[20] Lowry OH, Rosebrough NJ, Farr AL and Randall RJ 1951 Protein measurement with the Folin phenol reagent. *J. Biol. Chem.* **193** 265–275
[21] Nandini KE and Rastogi NK 2010 Single step purification of lactoperoxidase from whey

involving reverse micelles-assisted extraction and its comparison with reverse micellar extraction. *Biotechnol. Prog.* **26** 763–771

[22] He S, Shi J, Walid E, Ma Y and Xue SJ 2013 Extraction and purification of a lectin from small black kidney bean (*Phaseolus vulgaris*) using a reversed micellar system. *Process Biochem.* **48** 746–752

[23] Zhao X, Zhang X, Liu H and Sun H 2018 Optimization of AOT reversed micelle forward extraction of 7S globulin subunits from soybean proteins. *J. Food Sci. Technol.* **55** 4909–4917

[24] Pawar SS, Iyyaswami R and Belur PD 2017 Reverse micellar extraction of lactoferrin from its synthetic solution using CTAB/n-heptanol system. *J. Food Sci. Technol.* **54** 3630–3639

Acknowledgements

The authors acknowledged the financial support provided by Universiti Malaysia Pahang (UMP) through a short-term internal research grant (RDU170338) and a postgraduate research grant (PGRS170346).

AUTHOR INDEX